Holland, Richard
Design of resonant
piezoelectric devices

DESIGN OF RESONANT
PIEZOELECTRIC DEVICES

DESIGN OF RESONANT PIEZOELECTRIC DEVICES

RICHARD HOLLAND
E. P. EerNISSE

RESEARCH MONOGRAPH No. 56
THE M.I.T. PRESS
CAMBRIDGE, MASSACHUSETTS, AND
LONDON, ENGLAND

Set in Monotype Times Roman.
Printed and bound in the United States of America
by The Colonial Press Inc.

SBN 262 08033 8

Library of Congress catalog card number: 76–78096

Foreword

This is the fifty-sixth volume in the M.I.T. Research Monograph Series published by the M.I.T. Press. The objective of this series is to contribute to the professional literature a number of significant pieces of research, larger in scope than journal articles but normally less ambitious than finished books. We believe that such studies deserve a wider circulation than can be accomplished by informal channels, and we hope that this form of publication will make them readily accessible to research organizations, libraries, and independent workers.

HOWARD W. JOHNSON

Acknowledgment

This work was supported by the United States Atomic Energy Commission and by a National Science Foundation fellowship. Chapter 2 is part of a doctoral thesis submitted to the Department of Electrical Engineering, Massachusetts Institute of Technology, Cambridge, Mass., in January 1966.

Preface

This work was begun in June 1964, when the first author was a summer employee at Sandia Corporation, Albuquerque, New Mexico. At that time, Sandia was engaged in a research program dealing with piezoelectric ceramic devices. The goal of the program was to replace as much semiconductor circuitry as possible with piezoelectric counterparts. By doing this, the Sandia administration believed that the radiation tolerance of various systems could be greatly enhanced without a significant increase in component size or weight.

It was soon apparent, however, that in the field of piezoelectricity the mathematical techniques of 1964 were very inadequate for designing complex devices. Because of this, Sandia gave considerable priority to the development of more powerful analytical procedures. At this time, the first author had just completed a course in theoretical physics, given by Prof. H. Feshbach at M.I.T. He, consequently, was in the fortunate position of having recent familiarity with some very relevant mathematics.

The first new theoretical design approach which yielded real success was the Green's function and eigenmode technique described here in Chapter 2. This technique later became the subject of the first author's doctoral thesis, which was completed and submitted to the Department of Electrical Engineering at M.I.T. in January 1966. The remarkably simple nature of piezoelectric drive, which was first deduced at this time, is almost invariably met with incredulity by a reader until he has himself checked out a few equations.

Later in 1966, the second author joined in this research. The Lagrangian formulation reported here in Chapters 3 and 4 represents a mutual effort of the two authors. This formulation is an extension of a classical technique to a new problem class. The Lagrangian procedure is more tedious to apply than the Green's function procedure. However, the Lagrangian procedure is also more general in that with it any and all problems of linear piezoelectricity can be treated. The Green's function procedure on the other hand, is only applicable when one can make the approximation that elastic processes are one or two dimensional. Also, the Green's function procedure requires pre-knowledge of the system eigenmodes.

Chapter 5 describes some additional techniques for designing and understanding a particular class of devices which has recently gained much importance. These are the acoustic "energy-trapping" devices, which bear a distinct conceptual resemblance to electromagnetic waveguides.

Considerable vital support of technical or administrative nature has been given this project at Sandia by C. E. Land, I. D. McKinney, A. L. Roark, H. G. Baerwald, O. M. Stuetzer, and R. S. Claassen. The authors are also greatly indebted to Prof. J. R. Melcher at M.I.T. who was the first author's thesis advisor and who first suggested publication of this monograph.

Albuquerque, New Mexico
December 1968

RICHARD HOLLAND
E. P. EERNISSE

Contents

DESIGN OF RESONANT
PIEZOELECTRIC DEVICES

1 Introduction

The past eight or ten years have witnessed spectacular growth in piezoelectric device technology. During this period, both the sophistication and the variety of piezoelectric components in development or production have increased remarkably.

For example, while piezoelectric filters have been available for a long time, their frequency range has recently been greatly extended at both ends of the spectrum so that piezoelectric filters are now built from 1 kHz[1] to 200 MHz.[2]

Tranducers are perhaps the most obvious application for piezoelectric materials, as these devices take direct advantage of the defining characteristic of piezoelectricity — the coupling of mechanical and electrical effects. It is, of course, not possible at this point to discuss recent progress in the area of transducer technology. This area encompasses such diverse and rapidly moving fields as ultrasonic cleaning, microwave acoustic generation, sonar transmission and reception, and time delay.

In addition to the traditional applications of piezoelectrics in filters and transducers, a number of entirely new applications recently have been explored and developed. These new applications include electro-optic and elastooptic devices,[3] memory devices,[4,5] logic elements,[5] FM discriminators,[6] electrically tunable oscillators,[6] and high-precision thermometers.[7] It even appears possible that acoustic amplifiers utilizing piezoelectric semiconductor traveling wave elements could be manufactured at some time in the future when certain material reliability problems have been solved.[8,9]

From the devices mentioned so far, we observe a certain tendency to develop piezoelectric components that duplicate the function of already existing conventional components. There are two strong motivating reasons for this.

The first pertains to national defense and space applications: Piezoelectric miniaturized components are better able to survive radiation than conventional semiconductor microcircuitry. Hence, it is desirable to use piezoelectric circuitry extensively in any system that may be exposed either to a nuclear blast or to radiation over extended periods in space.

Second, it is often simpler to miniaturize a given component or group of components piezoelectrically than in any other way. A good example of this situation is found in the miniaturization of narrow band filters. Miniaturizing a filter component by component requires the availability of miniature high-Q inductors that are impossible to produce. (It can be shown that if an inductor is scaled down in size, the Q of the inductor will be reduced by the square of the scale factor.)[10] The usual alternative is to use active filters, but a high-Q active filter is, of course, beset with very difficult stability requirements. Consequently, this is a natural application for piezoelectric resonators that are passive and hence inherently stable but also have a very high Q that is independent of physical size or scale factor.

Along with these earlier-mentioned developments in actual hardware, a number of significant theoretical results have been obtained recently in the macroscopic behavior of piezoelectric devices. These include the Lagrangian and Green's function formulations of piezoelectricity and also the idea of "energy trapping." In many cases, these theoretical concepts permit rather clear and vivid visualizations of the complicated processes they model. Consequently, we feel they should be of value not only in providing concise design formulas and criteria for the piezoelectric engineer but should also sharpen the device designer's intuition and insight for piezoelectric phenomena.

We think it unfortunate that the recent developments in piezoelectric device technology have often been performed with the theoretical tools of the 1940's. This is especially regrettable when so much better mathematical equipment is now available. We hope to alter this situation at least slightly. Consequently, while this book is to be primarily a discussion and description of theoretical techniques, we shall often indicate how an engineer would apply these techniques to real design problems.

A final comment: the reader will soon observe that many of our results pertain to ferroelectric ceramics. There is a definite reason for this. The primary goal of this work is to introduce and explicate *concepts* not specific results. Consequently, we have deliberately chosen the piezoelectric material symmetry class that is the simplest to represent analytically. This minimizes the risk of getting hopelessly lost in details. In almost every case where we restrict our results to ferroelectric ceramics, the general concepts are in no way so restricted. The reader who wishes to apply these techniques to more complicated materials should be able to carry out the required generalizations on his own.

1.1. Definitions and Conventions

Throughout this work, our notation is chosen to conform closely to that set forth in the 1949 IRE Standards on Piezoelectric Crystals.[11] These Standards embrace rationalized MKS units specially adapted to describe piezoelectric processes. Our results will be limited to linear, although not necessarily lossless, effects and will predominately relate to sinusoidal steady state.

In particular, according to IRE Standards, the following definitions are applicable in the description of any linear piezoelectric material:

$$u_i = \text{particle displacement} \tag{1.1}$$

$$S_{ij} = \text{strain tensor}$$
$$= \tfrac{1}{2}(u_{i,j} + u_{j,i}) \tag{1.2}$$

$$T_{ij} = \text{stress tensor} \tag{1.3}$$

$$\phi = \text{electric potential} \tag{1.4}$$

$$E_m = \text{electric field}$$
$$= -\phi_{,m} \tag{1.5}$$

$$D_n = \text{electric displacement} \tag{1.6}$$

$$s_{ijkl}^E = \left.\frac{\partial S_{ij}}{\partial T_{kl}}\right|_{E=\text{const}} = \begin{array}{c}\text{adiabatic compliance tensor at}\\ \text{constant electric field}\end{array} \tag{1.7}$$

$$\epsilon_{mn}^T = \left.\frac{\partial D_m}{\partial E_n}\right|_{T=\text{const}} = \begin{array}{c}\text{adiabatic permittivity tensor}\\ \text{at constant stress}\end{array} \tag{1.8}$$

$$d_{njk} = \left.\frac{\partial D_n}{\partial T_{jk}}\right|_{E=\text{const}} = \left.\frac{\partial S_{jk}}{\partial E_n}\right|_{T=\text{const}}$$
$$= \text{adiabatic piezoelectric tensor} \tag{1.9}$$

In Equations 1.2 and 1.5 and hereafter, commas denote differentiation with respect to the following indices.

If the material represented by these coefficient tensors is assumed to be lossless, the equality of the two partial derivatives in Equation 1.9 is one of the so-called thermodynamic Maxwell relations. It may be established by defining the Gibbs function

$$G = U - S_{ij}T_{ij} - E_m D_m \tag{1.10}$$

and taking appropriate second derivatives.[12] (Repeated indices are understood to imply summation here and subsequently.) In Equation 1.10, U is the internal energy

$$dU = T_{ij}\, dS_{ij} + E_m\, dD_m \tag{1.11}$$

The two partial derivatives of Equation 1.9 are still undoubtedly equal if losses are present, but there is apparently no easy way to prove this statement.

In terms of the quantities described by Equations 1.1 to 1.9, it is possible to write the constituent relations for a general linear piezoelectric material:

$$\begin{aligned} S_{ij} &= s^E_{ijkl}T_{kl} + d_{mij}E_m \\ D_n &= d_{nkl}T_{kl} + \epsilon^T_{mn}E_m \end{aligned} \tag{1.12}$$

These relations show that piezoelectric substances differ from ordinary substances in that elastic and dielectric linear effects are coupled to each other. If the piezoelectric tensor d is set equal to zero, Equation 1.12 will reduce to the familiar generalized Hooke's law and dielectric equation.

In addition to the tensor method just discussed for describing piezoelectric processes, a second method is in common use and will often be helpful to us. This is the matrix or engineering method. Tensor notation is characterized by subscripts that run from 1 to 3, and which are usually represented by Latin letters. However, a certain redundancy is inherent in tensor subscripts, as

$$\begin{aligned} S_{ij} &= S_{ji} \qquad s^E_{ijkl} = s^E_{jikl} = s^E_{klij} \\ T_{ij} &= T_{ji}\dagger \qquad d_{mij} = d_{mji} \end{aligned} \tag{1.13}$$

Engineering notation takes advantage of this symmetry by incorporating two Latin indices into a single Greek index running from 1 to 6.

† Volume torque, proportional to **P** × **E**, may always be neglected in piezoelectric and ferroelectric materials.

The convention relating engineering and tensor indices is

$$\begin{array}{ll} ij = ji \rightarrow \nu & \\ 11 \rightarrow 1 & 23 = 32 \rightarrow 4 \\ 22 \rightarrow 2 & 13 = 31 \rightarrow 5 \\ 33 \rightarrow 3 & 12 = 21 \rightarrow 6 \end{array} \tag{1.14}$$

Certain other changes are involved in the transition from tensor to engineering notation.[11,13] Stress goes over without modification,

$$T_{ij} = T_\nu \tag{1.15}$$

Strain does not, however. It is desired to be able to write internal energy analogously in the two systems:

$$\begin{aligned} dU &= T_{ij}\, dS_{ij} + E_m\, dD_m \\ &= T_\nu\, dS_\nu + E_m\, dD_m \end{aligned} \tag{1.16}$$

The first form contains expressions such as $T_{12}\, dS_{12} + T_{21}\, dS_{21}$, while the second form only contains expressions such as $T_6\, dS_6$. It is consequently necessary to define engineering strain with a factor of 2 added in the shear components:

$$S_{ij} = \theta(\nu) S_\nu \qquad \theta(\nu) = \begin{cases} \nu = 1, 2, 3 \\ \frac{1}{2}\nu = 4, 5, 6 \end{cases} \tag{1.17}$$

Finally, it is necessary to relate the engineering compliance and piezoelectric coefficients to the corresponding tensor quantities. In engineering notation, the constituent relations are expressed

$$\begin{aligned} S_\nu &= s^E_{\nu\mu} T_\mu + d_{m\nu} E_m \\ D_n &= d_{n\mu} T_\mu + \epsilon^T_{nm} E_m \end{aligned} \tag{1.18}$$

The consistency of Equation 1.18 with Equations 1.12 then requires

$$\begin{aligned} d_{mkl} &= \theta(\nu)\, d_{m\nu} \\ s^E_{ijkl} &= \theta(\mu)\, \theta(\nu) s^E_{\nu\mu} \end{aligned} \tag{1.19}$$

This just-described engineering notation is often alternatively called matrix notation because Equations 1.18, the engineering constituent relations, can so easily be represented by matrices:

$$\begin{aligned} [S] &= [s^E][T] + [d_t][E] \\ [D] &= [d][T] + [\epsilon^T][E] \end{aligned} \tag{1.20}$$

Here, d_t is the transposed d array.

It is, of course, possible to write the constituent equations with variables other than $[T]$ and $[E]$ independent. For instance, if we solve Equations 1.20 for $[S]$ and $[E]$, we find

$$\begin{aligned} [S] &= [s^D][T] + [g_t][D] \\ [E] &= -[g][T] + [\beta^T][D] \end{aligned} \tag{1.21}$$

where

$$\begin{aligned} [\beta^T] &= [\epsilon^T]^{-1} \\ [g] &= [\beta^T][d] \\ [s^D] &= [s^E] - [d_t][\beta^T][d] \end{aligned} \tag{1.22}$$

If Equations 1.20 are solved for $[T]$ and $[D]$, we find

$$\begin{aligned} [T] &= [c^E][S] - [e_t][E] \\ [D] &= [e][S] + [\epsilon^S][E] \end{aligned} \tag{1.23}$$

where

$$\begin{aligned} [c^E] &= [s^E]^{-1} \\ [e] &= [d][c^E] \\ [\epsilon^S] &= [\epsilon^T] - [d][c^E][d_t] \end{aligned} \tag{1.24}$$

It is also possible to write the constituent relations with $[S]$ and $[D]$ independent:

$$\begin{aligned} [T] &= [c^D][S] - [h_t][D] \\ [E] &= -[h][S] + [\beta^S][D] \end{aligned} \tag{1.25}$$

where

$$\begin{aligned} [c^D] = [s^D]^{-1} &= [c^E] + [e_t][\beta^S][e] \\ &= [c^E] + [h_t][\epsilon^S][h] \\ [h] = [\beta^S][e] &= [g][c^D] \\ [\beta^S] = [\epsilon^S]^{-1} &= [\beta^T] + [h][s^D][h_t] \\ &= [\beta^T] + [g][c^D][g_t] \end{aligned} \tag{1.26}$$

Finally, it is possible to transform the matrix relations, Equations 1.21, 1.23, and 1.25, back into tensor notation in order to obtain tensor constituent relations with variables other than T_{kl} and E_m independent.

This operation leads to

$$\begin{aligned} S_{ij} &= s^D_{ijkl} T_{kl} + g_{mij} D_m \\ E_n &= -g_{nkl} T_{kl} + \beta^T_{nm} D_m \end{aligned} \tag{1.27}$$

$$\begin{aligned} T_{ij} &= c^E_{ijkl} S_{kl} - e_{mij} E_m \\ D_n &= e_{nkl} S_{kl} + \epsilon^S_{nm} E_m \end{aligned} \tag{1.28}$$

$$\begin{aligned} T_{ij} &= c^D_{ijkl} S_{kl} - h_{mij} E_m \\ E_n &= -h_{nkl} S_{kl} + \beta^S_{nm} E_m \end{aligned} \tag{1.29}$$

The tensor and engineering coefficients appearing here are related to each other by

$$\begin{aligned} s^{E,D}_{ijkl} &= \theta(\mu)\theta(\nu) s^{E,D}_{\mu\nu} \\ c^{E,D}_{ijkl} &= c^{E,D}_{\mu\nu} \\ d_{mjk} &= \theta(\nu)\, d_{m\nu} \qquad g_{mjk} = \theta(\nu) g_{m\nu} \\ e_{mjk} &= e_{m\nu} \qquad h_{mjk} = h_{m\nu} \end{aligned} \tag{1.30}$$

As a general rule, tensor notation is most useful in problems involving coordinate transformations or lengthy derivations, and matrix notation is most useful for carrying out numerical computations pertaining to specific problems.

1.2. Examples of Piezoelectric Crystal Coefficient Arrays

All piezoelectric materials lack a center of symmetry.[14] Hence piezoelectricity is inherently an anisotropic phenomenon. Moreover, piezoelectricity is so intimately related to elastic and dielectric effects that it is almost nonsensical to discuss the piezoelectric properties of some material without considering these other properties as well.

Perhaps the most concise method of representing the electroelastic properties of a substance is that originated by K. S. Van Dyke.[15] Basically there are 32 classes of crystals, 20 of which lack a center of symmetry and thus may be piezoelectric. Diagrams representing the general forms of the electroelastic coefficient arrays for each of these 32 classes according to the Van Dyke scheme are given in Table 1.1. In each block of this table, the large 6 × 6 array of dots or spaces in the upper left corner represents the elastic coefficient matrix ($c^E_{\mu\nu}$, $c^D_{\mu\nu}$, $s^E_{\mu\nu}$, or $s^D_{\mu\nu}$). The small 3 × 3 arrays in the lower right corners correspond to the dielectric coefficient matrices (ϵ^S_{mn}, ϵ^T_{mn}, β^S_{mn}, or β^T_{mn}). The

Table 1.1.* Elasto-Piezo-Dielectric Matrix of Various Crystal Classes

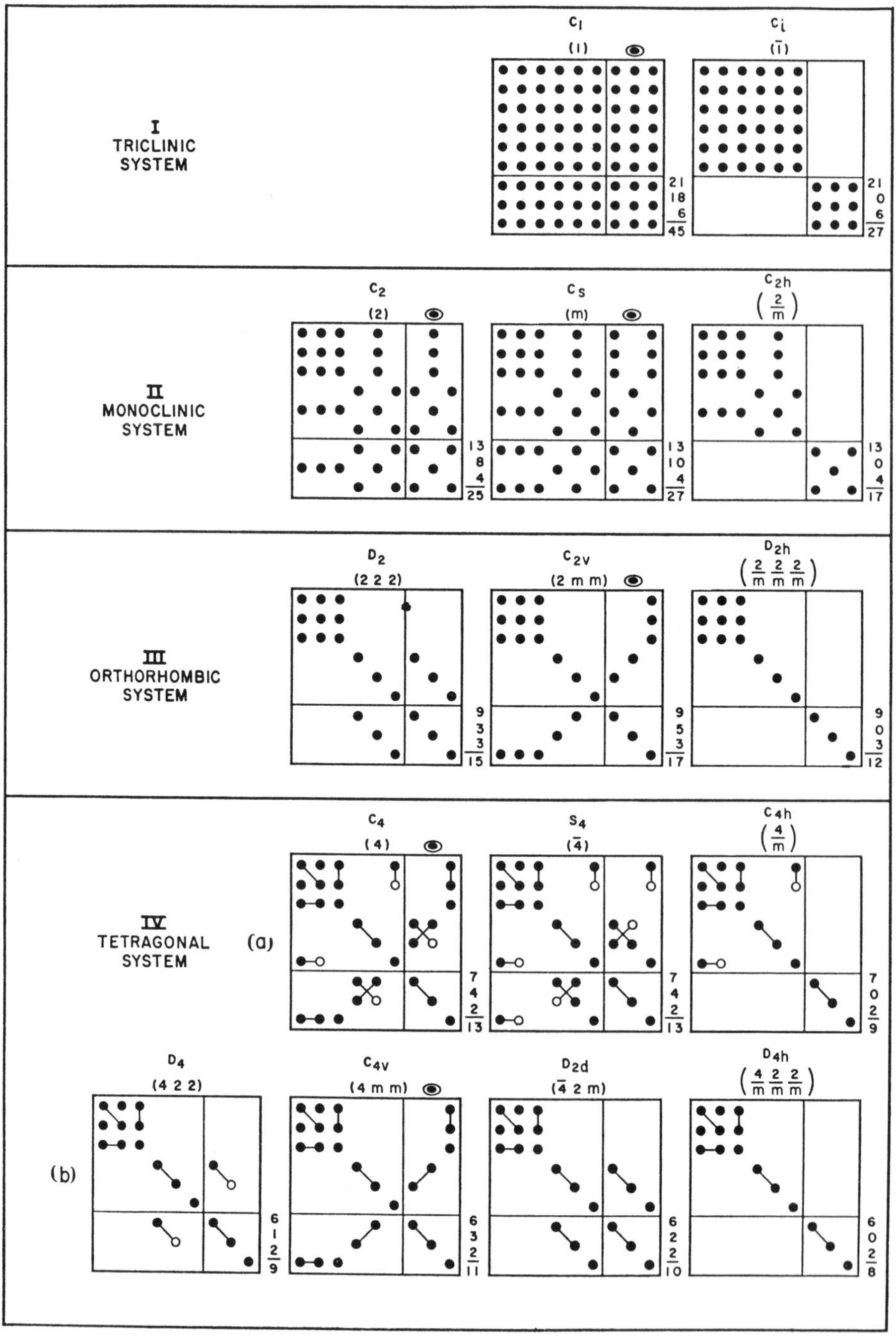

* The numbers on the right side of each scheme indicate, from top to bottom, the number of the independent elastic, piezoelectric, and dielectric constants. Reprinted from *Proc. IRE*, vol. 46, pp. 767 to 768.

Table 1.1 (continued)

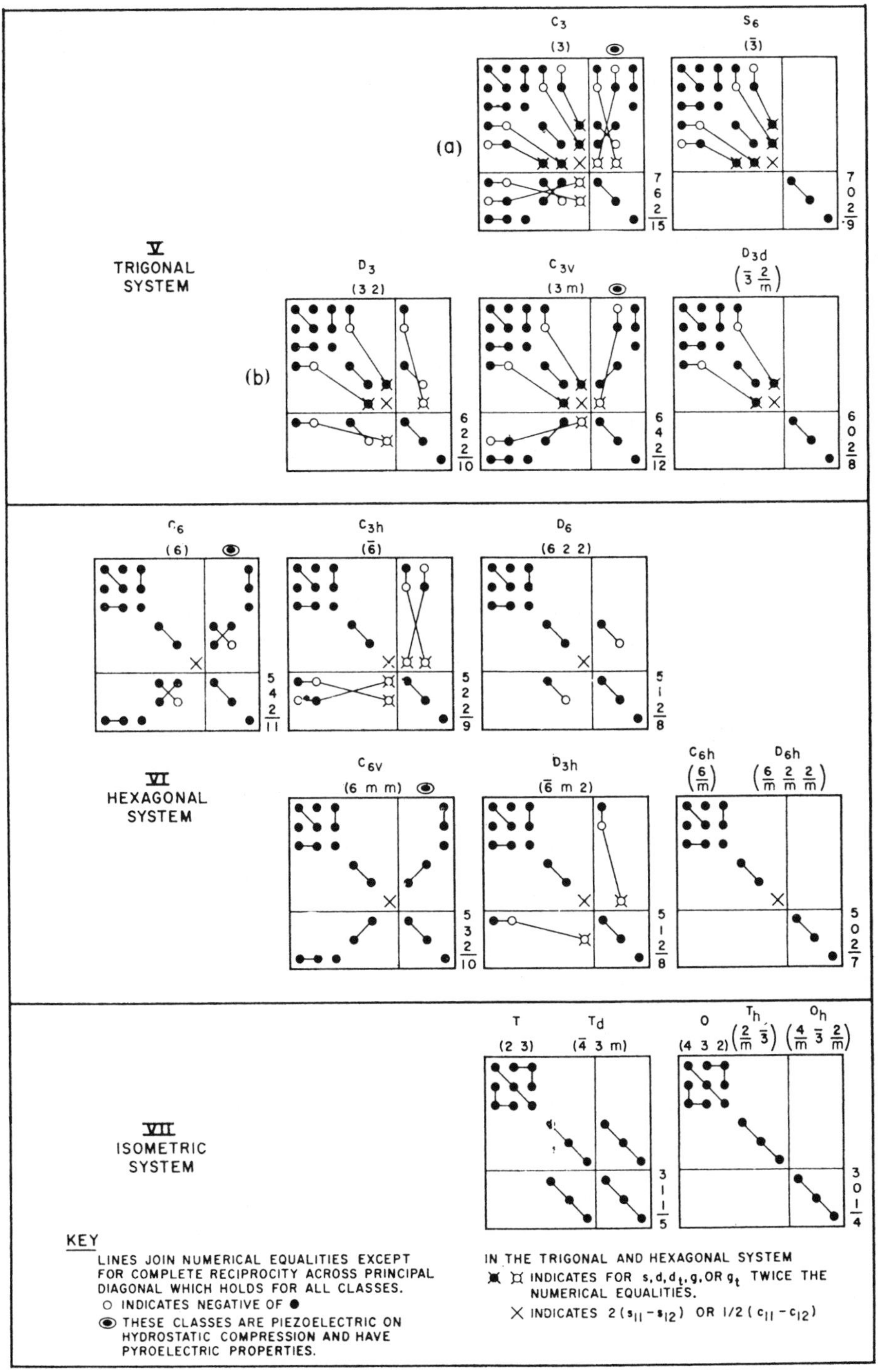

two remaining 3 × 6 and 6 × 3 arrays represent the piezoelectric matrix and its transpose ($d_{m\nu}$, $e_{m\nu}$, $g_{m\nu}$, or $h_{m\nu}$). The crystal class represented by each of these blocks is shown above the block, both in Schoenfleiss and Hermann-Maughm notation.

Blank spaces in these diagrams indicate that the coefficient corresponding to that space is zero for the class in question. Solid dots represent nonzero coefficients; connected dots represent equality between nonzero coefficients. Open circles connected to solid dots indicate one coefficient is the negative of the other. Additional explanations of these tables may be found at the bottom of the second page of the table.

At least one type of material, ferroelectric ceramic, is piezoelectric even though it is not a single crystal. However, ferroelectric ceramics have coefficient arrays of the same form as hexagonal crystal class $C_{6\nu}$. These ceramics, in fact, are very strongly piezoactive—more so than any nonferroelectric single crystal. This is one reason they are very interesting both from the viewpoints of application and of behavioral theory.

In concluding this section, it is convenient to give numerical values for the electroelastic coefficients of some representative members of four crystal or material classes. We shall begin with quartz, which is probably still the most widely used piezoelectric material, especially at high frequencies. Table 1.2 presents the coefficients for quartz (class D_3) in the $s^E_{\mu\nu}$, $d_{m\nu}$, ϵ^T_{mn} formulation.[16]

Table 1.2. Electroelastic Coefficients for Quartz (MKS units)[16]

$s^E_{11} = 12.77 \times 10^{-12}$	$d_{11} = 2.31 \times 10^{-12}$
$s^E_{33} = 9.60 \times 10^{-12}$	$d_{14} = 0.727 \times 10^{-12}$
$s^E_{12} = -1.79 \times 10^{-12}$	$\epsilon^T_{11} = 4.52\epsilon_0$
$s^E_{13} = -1.22 \times 10^{-12}$	$\epsilon^T_{33} = 4.68\epsilon_0$
$s^E_{44} = 20.04 \times 10^{-12}$	
$s^E_{66} = 29.12 \times 10^{-12}$	
$s^E_{14} = 4.50 \times 10^{-12}$	

Table 1.3 gives the corresponding coefficients for the ferroelectric ceramic PZT-5*,[17] which is a lead-zirconate-titanate composition. It is interesting to compare how large the piezoelectric coefficients for PZT-5 are in relation to those of quartz crystals. PZT-5 is one of the most strongly piezoelectric substances yet developed.

* Clevite Corporation, trade name.

Table 1.3. Electroelastic Coefficients for the Lead-Zirconate-Titanate Ceramic PZT-5 (MKS units)[17]

$s_{11}^E = 16.4 \times 10^{-12}$	$d_{31} = -171 \times 10^{-12}$
$s_{33}^E = 18.8 \times 10^{-12}$	$d_{33} = 374 \times 10^{-12}$
$s_{12}^E = -5.74 \times 10^{-12}$	$d_{15} = 584 \times 10^{-12}$
$s_{13}^E = -7.72 \times 10^{-12}$	$\epsilon_{11}^T = 1730\epsilon_0$
$s_{44}^E = 47.5 \times 10^{-12}$	$\epsilon_{33}^T = 1700\epsilon_0$
$s_{66}^E = 44.3 \times 10^{-12}$	

Lithium niobate (class C_{3v}) is one of the more recently developed piezoelectric single crystals. This material compares favorably with the ferroelectric ceramics in that its piezoelectric effects are nearly as strong and its losses are much lower than those of the ceramics. The independent electroelastic parameters for lithium niobate are given in Table 1.4.[18]

Table 1.4. Electroelastic Coefficients for Lithium Niobate, $LiNbO_3$ (MKS units)[18]

$s_{11}^E = 5.78 \times 10^{-12}$	$d_{15} = 68 \times 10^{-12}$
$s_{33}^E = 5.02 \times 10^{-12}$	$d_{22} = 21 \times 10^{-12}$
$s_{12}^E = -1.01 \times 10^{-12}$	$d_{31} = -1 \times 10^{-12}$
$s_{13}^E = -1.47 \times 10^{-12}$	$d_{33} = 6 \times 10^{-12}$
$s_{44}^E = 17.0 \times 10^{-12}$	$\epsilon_{11}^T = 84\epsilon_0$
$s_{66}^E = 13.6 \times 10^{-12}$	$\epsilon_{33}^T = 30\epsilon_0$
$s_{14}^E = -1.02 \times 10^{-12}$	

Bismuth germanium oxide, $Bi_{12}GeO_{20}$ (class T), is also a recently developed piezoelectric single crystal. It is characterized by relatively low losses, especially at microwave frequencies, but yet is considerably more strongly piezoactive than quartz. Consequently, it may prove to be quite useful in microwave acoustic devices. This is the only known crystal in class T which is insoluble in water and is refractory. Class T is, in turn, the only cubic class which is both piezoelectrically and optically active. The independent electroelastic coefficients of bismuth germanium oxide are given in Table 1.5.[19,20]

Table 1.5. Electroelastic Coefficients for Bismuth Germanium Oxide, $Bi_{12}GeO_{20}$ (MKS units)[19]

$s_{11}^E = 9.9 \times 10^{-12}$	$d_{14} = 28 \times 10^{-12}$
$s_{12}^E = -2.43 \times 10^{-12}$	$\epsilon_{11}^T = 40\epsilon_0$
$s_{44}^E = 40.0 \times 10^{-12}$	

1.3. Representation of Loss

Throughout this monograph, we shall be primarily concerned with the behavior of nonideal materials in sinusoidal steady state. It is thus necessary to be able to represent material loss under this type of excitation.

The use of a complex dielectric constant or loss tangent to represent dielectric dissipation phenomenologically is well known as is the use of complex compliance to represent elastic relaxation. Martin[21,22] and Land[23] have suggested extending this concept to piezoelectric coefficients to describe imperfect energy conversion. This approach proves to be quite useful for our purposes in the work ahead. For this reason we will now take some time to explore a few of the consequent physical and mathematical implications of this concept.

Our starting point will be the piezoelectric constituent relations in tensor form, Equation 1.12,

$$\begin{aligned} S_{ij} &= s^E_{ijkl} T_{kl} + d_{mij} E_m \\ D_n &= d_{nkl} T_{kl} + \epsilon^T_{mn} E_m \end{aligned} \tag{1.31}$$

To this is added the equation of magnetization for nonmagnetic materials,

$$B_m = \mu_0 H_m \tag{1.32}$$

All field and displacement quantities in the present section contain a suppressed $e^{j\omega t}$ dependence.

Material coefficients (excluding μ_0) are to be regarded here as complex. It will develop shortly that positive loss corresponds to negative phase angles in these coefficients. Thus it is simplest if we define the sign of the imaginary parts of s^E_{ijkl}, d_{mij}, and ϵ^T_{mn} by

$$\begin{aligned} s^E_{ijkl} &= s^{E'}_{ijkl} - j s^{E''}_{ijkl} \\ d_{mij} &= d'_{mij} - j d''_{mij} \\ \epsilon^T_{mn} &= \epsilon^{T'}_{mn} - j \epsilon^{T''}_{mn} \end{aligned} \tag{1.33}$$

It is possible to form a generalized energy flux or Poynting vector in a piezoelectric material as the sum of the electrical and mechanical energy fluxes,

$$\Sigma_i = \tfrac{1}{2}[(\mathbf{E} \times \mathbf{H}^*)_i - T_{ij} \dot{u}_j^*] \tag{1.34}$$

In this equation, an asterisk denotes complex conjugation and $\dot{u}_j$ is the particle velocity or time derivative of u_j.

The negative real part of the divergence of this vector is the power dissipation density in the material,

$$\begin{aligned} P_d &= -\mathrm{Re}\,(\Sigma_{i,i}) \\ &= -\tfrac{1}{2}\,\mathrm{Re}\,[H_i^*(\nabla \times \mathbf{E})_i - E_i(\nabla \times \mathbf{H}^*)_i \\ &\quad + j\omega T_{ij}u_{i,j}^* + j\omega u_j^* T_{ij,i}] \end{aligned} \tag{1.35}$$

Tensor identities necessary to derive this relation may be obtained from Morse and Feshbach.[24] We shall now proceed to simplify this expression for P_d in such a way as to give a straightforward physical interpretation to electroelastic phase angles.

Maxwell's equations in a nonconducting medium are

$$(\nabla \times \mathbf{E})_i = -j\omega B_i \qquad (\nabla \times \mathbf{H})_i = j\omega D_i \tag{1.36}$$

The double contraction of a symmetric tensor such as T_{ij} with the antisymmetric part of $u_{i,j}$ is zero. As strain is the symmetric part of $u_{i,j}$, we thus obtain

$$T_{ij}u_{i,j} = T_{ij}S_{ij} \tag{1.37}$$

Newton's equation of momentum conservation for a distributed medium is

$$T_{ij,i} = -\omega^2 \rho u_j \tag{1.38}$$

where ρ is the material density. If Equations 1.31, 1.32, and 1.36 to 1.38 are used to simplify and reduce Equation 1.35, we obtain

$$P_d = (\tfrac{1}{2}\omega)\,\mathrm{Im}\,(E_m D_m^* + T_{ij}S_{ij}^*) \tag{1.39}$$

At this point it is convenient to convert to engineering notation and to substitute Equation 1.18, the engineering constituent relations. These operations lead to expressing the power dissipation density as

$$P_d = (\tfrac{1}{2}\omega)\,\mathrm{Im}\,(E_m d_{m\nu}^* T_\nu^* + E_m \epsilon_{mn}^{T*} E_n^* + T_\mu s_{\mu\nu}^{E*} T_\nu^* + T_\mu d_{m\mu}^* E_m^*) \tag{1.40}$$

Let us now define a 9-dimensional generalized force vector F_A

$$[F_A] = \begin{bmatrix} T_\nu \\ E_m \end{bmatrix} \tag{1.41}$$

and a 9 × 9 symmetric coefficient matrix

$$[M_{AB}] = \begin{bmatrix} s_{\mu\nu}^E & d_{\nu m} \\ d_{m\nu} & \epsilon_{mn}^T \end{bmatrix} \tag{1.42}$$

where $d_{\nu m}$ is the transpose of $d_{m\nu}$. The subscripts A and B are understood to run 1 to 9. In terms of F_A and M_{AB}, P_d is reduced to

$$P_d = (\tfrac{1}{2}\omega)\ \mathrm{Im}\ (F_A M^*_{AB} F^*_B) \tag{1.43}$$

If we let θ_{AB} be the phase angle between F_A and F_B, we can further simplify this relation:

$$\begin{aligned} P_d &= (\tfrac{1}{2}\omega) \sum_{A,B=1}^{9} |F_A|\,|F_B|\,(M''_{AB}\cos\theta_{AB} - M'_{AB}\sin\theta_{AB}) \\ &= (\tfrac{1}{2}\omega) \sum_{A,B=1}^{9} |F_A|\,|F_B|\,M''_{AB}\cos\theta_{AB} \end{aligned} \tag{1.44}$$

The sin θ_{AB} terms drop out of this equation because sin θ_{BA} cancels sin θ_{AB}. In Equation 1.44, the repeated index summation convention is dropped, $|F_A|$ is defined as the absolute value of the complex quantity F_A, and M''_{AB} is defined in analogy with Equation 1.33,

$$M_{AB} = M'_{AB} - jM''_{AB} \tag{1.45}$$

Let us now leave off this mathematical discussion and take stock of the physical implications of our results so far. First, in view of the definitions of F_A and M_{AB}, we see that $s^{E''}_{\mu\nu}$ is associated with power dissipation of a purely mechanical nature of the form $\frac{1}{2}\omega|T_\mu|\,|T_\nu|s^{E''}_{\mu\nu}\cos\theta_{\mu\nu}$. Similarly, $\epsilon^{T''}_{mn}$ is associated with a purely electrical power loss, $\frac{1}{2}\omega|E_m|\,|E_n|\epsilon^{T''}_{mn}\cos\theta_{mn}$. In addition, $d''_{m\mu}$ is associated with a third type of power loss, $\frac{1}{2}\omega|E_m|\,|T_\mu|d''_{m\mu}\cos\theta_{m\mu}$, which is truly electromechanical, or dependent on both E_m and T_μ.

It is interesting to observe that off-diagonal components of M_{AB} correspond to power *gain*, not loss, if M''_{AB} is positive and θ_{AB} is not between $-\pi/2$ and $\pi/2$. We shall shortly see that this curious result puts a number of constraints on the permitted values of M''_{AB} in a passive material.

For example, the nine components of F_A are all independent variables. Thus it is possible to pick them so $\theta_{AB} = 0$ or π for all A and B. If this is done, P_d may be written as

$$P_d = (\tfrac{1}{2}\omega)F_A M''_{AB} F_B \tag{1.46}$$

where all the F_A are real although not necessarily positive. For a passive material, P_d can never be negative, regardless of the choice of F_A. This statement and Equation 1.46 imply that the matrix M''_{AB} must be positive definite.[25] We shall now see just how this requirement restricts the material parameters.

It is a theorem of linear algebra that all the principal minor determinants of a positive definite matrix must be positive.[26] An mth-order principal minor determinant of an nth-order matrix is obtained as follows: First cross out any $n - m$ rows of the matrix. Then cross out the $n - m$ columns specified by reflecting the crossed-out rows across the main diagonal of the original matrix. Finally, take the determinant of the remaining mth-order matrix. It can also be shown that an nth-order matrix has $2^n - 1$ principal minor determinants. Thus M''_{AB} will have 511 principal minor determinants, all of which must be positive. Consequently, the requirement that a material be passive puts 511 constraints on the permitted values of M''_{AB}, although, at most, nine of these constraints are independent.

For example, the positiveness of the first-order principal minor determinants requires

$$M''_{AA} \geq 0 \tag{1.47}$$

where the summation convention is not invoked. Because of this, we chose the sign of the imaginary parts of the material coefficients to be negative as expressed in Equation 1.33. This inequality implies that for a material to be passive all the diagonal elements of the compliance and dielectric matrices must have negative phase angles.

The positiveness of the second- and third-order principal minor determinants, in turn, requires[26]

$$M''_{AA}M''_{BB} \geq (M''_{AB})^2 \tag{1.48}$$

$$[M''_{AA}M''_{BB} - (M''_{AB})^2][M''_{AA}M''_{CC} - (M''_{AC})^2] \geq (M''_{AA}M''_{BC} - M''_{AB}M''_{AC})^2 \tag{1.49}$$

Let us now conclude this topic with an actual numerical example. Apparently there is no piezoelectric material for which the real and imaginary parts of all the elements of M_{AB} are known. However, for a certain lead-zirconate-titanate hot-pressed ferroelectric ceramic $PbZr_{0.65}Ti_{0.35}O_3$ (PZT 65/35), enough of these properties are known to make some interesting observations. Ferroelectric ceramics are described by an M_{AB} matrix of the same form as crystals of class C_{6v}. In other words, the independent coefficients are s^E_{11}, s^E_{12}, s^E_{13}, s^E_{33}, s^E_{44}, d_{31}, d_{33}, d_{15}, ϵ^T_{11}, and ϵ^T_{33} (see Table 1.1). When Equations 1.47 and 1.48 are applied to these nine quantities, it is found

$$s^{E''}_{11}, s^{E''}_{33}, s^{E''}_{44}, \epsilon^{T''}_{11}, \epsilon^{T''}_{33} > 0 \tag{1.50}$$

and

$$\begin{aligned} s_{11}^{E''} &\geq |s_{12}^{E''}| && A = 1, \quad B = 2 \\ s_{11}^{E''} s_{33}^{E''} &\geq (s_{13}^{E''})^2 && A = 1, \quad B = 3 \\ s_{11}^{E''} \epsilon_{33}^{T''} &\geq (d_{31}'')^2 && A = 1, \quad B = 9 \\ s_{33}^{E''} \epsilon_{33}^{T''} &\geq (d_{33}'')^2 && A = 3, \quad B = 9 \\ s_{44}^{E''} \epsilon_{11}^{T''} &\geq (d_{15}'')^2 && A = 4, \quad B = 8 \end{aligned} \tag{1.51}$$

In addition, it can be shown that the positiveness of the (1,2,3) and (1,2,9) third-order principal minor determinants requires

$$\begin{aligned} s_{33}^{E''}(s_{11}^{E''} + s_{12}^{E''}) &\geq 2(s_{13}^{E''})^2 \\ \epsilon_{33}^{T''}(s_{11}^{E''} + s_{12}^{E''}) &\geq 2(d_{31}^{E''})^2 \end{aligned} \tag{1.52}$$

Measurements have been made on s_{11}^E, s_{12}^E, ϵ_{33}^T, and d_{31} for PZT 65/35; the results are reproduced in Table 1.6.[27] (The values in Table 1.6 are

Table 1.6. Complex Electroelastic Coefficients of PZT 65/35 (MKS units)[27]

$s_{11}^E = 9.2 \times 10^{-12} - j \times 0.057 \times 10^{-12}$
$s_{12}^E = -2.7 \times 10^{-12} + j \times 0.017 \times 10^{-12}$
$d_{31} = -43 \times 10^{-12} + j \times 1.4 \times 10^{-12}$
$\epsilon_{33}^T = 3900 \times 10^{-12} - j \times 130 \times 10^{-12} = (440 - j15)\epsilon_0$

for maximally polarized PZT 65/35. results for partially polarized samples are given in Reference 27.) We can observe that the first and third inequalities of Equation 1.51 are indeed obeyed by the coefficients given in Table 1.6.

The last three inequalities of Equation 1.51 lead to the curious theorem that (at least for a ferroelectric ceramic) if a material is either perfectly elastic or dielectric, the piezoelectric conversion properties will be perfect as well.

1.4. Stationary or Eigen Coupling Factors and Loss Tangents*

1.4.1. Eigen Coupling Factors; the Lossless DC Case

In many respects, it turns out that the piezoelectric constants themselves are not the best of all possible measures of the strength of electroelastic coupling in a given material. Usually the dimensionless coupling

* This section may be omitted without loss of continuity.

factor is actually a more useful parameter. This coupling factor is commonly defined as follows.

The energy storage density in a lossless piezoelectric material under linear dc conditions may, according to Equations 1.16 and 1.18, be represented as

$$\begin{aligned} 2U &= E_m \epsilon_{mn}^T E_n + 2E_m d_{m\nu} T_\nu + T_\mu s_{\mu\nu}^E T_\nu \\ &= 2(U_D + 2U_M + U_E) \end{aligned} \tag{1.53}$$

Here U_D is the dielectric energy, U_E is the elastic energy, and U_M is the energy of electromechanical coupling. The coupling factor is then given by [28,29]

$$k' = \frac{U_M}{\sqrt{U_E U_D}} \tag{1.54}$$

Alternatively, the coupling factor is occasionally defined by replacing the geometric mean in Equation 1.54 by the arithmetic mean,[30,31]

$$k = \frac{2U_M}{U_E + U_D} \tag{1.55}$$

Unfortunately, it is impossible at this point to give the uninitiated reader much insight into the utility of these definitions. For the time being, he will have to accept on faith our statement that they recur at many crucial crossroads in the theory of phenomenological piezoelectricity. Subsequent chapters of this monograph, for instance, will indicate that this is true.

In the discussion of coupling factors given in this section, we will primarily be concerned with a series of mathematical manipulations of definitions in Equations 1.54 and 1.55. These manipulations lead fairly quickly to a result involving the conditions under which k or k' as defined earlier is stationary with respect to small perturbations in T_ν or E_m. Determination of these conditions was originally performed by Baerwald, who used a derivation of considerable complexity.[30,31]

The discussion of this section is potentially of value in a practical sense, because it indicates how to apply a very general theorem pertaining to the design of maximum bandwidth filters. Later in this section that theorem (which oddly has apparently never been proved but has no known counterexamples) will be presented.

We begin the formal theory and derivations of this section by rearranging Equation 1.54 and differentiating it with respect to E_m

and T_ν:

$$U_M d_{m\nu} T_\nu = k'^2 U_E \epsilon_{mn}^T E_n + 2k' \frac{dk'}{dE_m} U_E U_D$$
$$U_M d_{m\nu} E_m = k'^2 U_D s_{\mu\nu}^E T_\mu + 2k' \frac{dk'}{dT_\nu} U_E U_D \tag{1.56}$$

Equations 1.54 and 1.56 may be combined to show

$$k' s_{\mu\nu}^E \left(\sqrt{\frac{U_M}{U_E}}\, T_\nu\right) - d_{m\mu}\left(\sqrt{\frac{U_M}{U_D}}\, E_m\right) = -2k' \frac{dk'}{dT_\mu} U_E \sqrt{\frac{U_D}{U_M}}$$
$$-\, d_{n\nu}\left(\sqrt{\frac{U_M}{U_E}}\, T_\nu\right) + k'\epsilon_{mn}^T\left(\sqrt{\frac{U_M}{U_D}}\, E_m\right) = -2k' \frac{dk'}{dE_n} U_D \sqrt{\frac{U_E}{U_M}} \tag{1.57}$$

For certain states of the electroelastic variables (T_ν, E_m), say $(T_\nu, E_m) = (T'_\nu, E'_m)$, the coupling factor k' is stationary with respect to small variations in T_ν and E_m. These states are called the coupling eigenstates of the material. In other words, if (T'_ν, E'_m) is a coupling eigenstate, the derivatives appearing on the right side of Equations 1.57 are zero when evaluated at (T'_ν, E'_m). Thus the determinant of the coefficient matrix of $(T_\nu\sqrt{U_M/U_E},\ E_m\sqrt{U_M/U_D})$ in Equations 1.57 is zero if k' is stationary:

$$\begin{vmatrix} -k' s_{\mu\nu}^E & d_{\nu m} \\ d_{n\nu} & -k'\epsilon_{nm}^T \end{vmatrix} = 0 \tag{1.58}$$

Equation 1.58 is ninth order in k', and thus has nine solutions. To each of the nine stationary or eigen coupling factors, there corresponds a coupling eigenstate (T'_ν, E'_m).

If, instead of starting with Equation 1.54, the geometric mean definition for k', one starts by rearranging Equation 1.55, the algebraic mean definition for k, and then differentiating with respect to E_m and T_ν, we have

$$k s_{\mu\nu}^E T_\nu - d_{m\mu} E_m = -\frac{dk}{dT_\mu}(U_D + U_E)$$
$$-d_{n\nu} T_\nu + k\epsilon_{nm}^T E_m = -\frac{dk}{dE_n}(U_D + U_E) \tag{1.59}$$

in place of Equation 1.57. Consequently, Equation 1.55 leads to eigen coupling factors k which are the same as the solutions of Equation 1.58 for k'. However, the coupling eigenstates associated with Equation 1.55

are different from those associated with Equation 1.54. In particular, if Equation 1.55 and Equations 1.59 lead to a coupling eigenstate (T_ν, E_m), then Equations 1.54 and 1.57 will lead to a corresponding state (T'_ν, E'_m), where

$$\sqrt{\frac{U_M(T'_\nu, E'_m)}{U_E(T'_\nu)}}\, T'_\nu = T_\nu$$
$$\sqrt{\frac{U_M(T'_\nu, E'_m)}{U_D(E'_m)}}\, E'_m = E_m \qquad (1.60)$$

While it is generally possible to find the unprimed eigenstates by conventional matrix operations, the determination of the primed eigenstates would have to proceed by first finding the unprimed states and then solving Equations 1.60. These nine simultaneous quadratic equations for (T'_ν, E'_m) will usually be very difficult to solve. Thus working with the algebraic mean definition for k has great computational advantages over the alternative formulation involving k' when one has need to evaluate the associated coupling eigenstates. Subsequently in this section, we shall assume the coupling factor is k, not k', unless an explicit statement to the contrary is made.

Let us now indicate how one actually goes about computing the eigen coupling factors and eigenstates and describe some of the mathematical properties of these quantities. In analogy with Equation 1.42, we define the 9 × 9 symmetric coefficient matrices

$$[M^0_{AB}] = \begin{bmatrix} s^E_{\mu\nu} & 0 \\ 0 & \epsilon^T_{mn} \end{bmatrix}$$
$$[M^1_{AB}] = \begin{bmatrix} 0 & d_{\mu m} \\ d_{n\nu} & 0 \end{bmatrix} \qquad (1.61)$$

where the indices A and B will again be assumed to run 1 to 9. In the future, these matrices may be abbreviated simply as $[M^0]$ and $[M^1]$. The equation for the eigen coupling factors k may then be written

$$|M^1_{AB} - kM^0_{AB}| = 0 \qquad (1.62)$$

We shall also utilize here the nine-dimensional generalized force of Equation 1.41,

$$[F_A] = \begin{bmatrix} T_\nu \\ E_m \end{bmatrix} \qquad (1.63)$$

which we may abbreviate as $[F]$.

In terms of Equations 1.61 and 1.63, it is possible to represent the energy components associated with $[F]$ by

$$\begin{aligned} 2U_M &= \tfrac{1}{2}[F_t][M^1][F] \\ U_E + U_D &= \tfrac{1}{2}[F_t][M^0][F] \end{aligned} \tag{1.64}$$

where a subscript t indicates the transpose of a vector or matrix.

Let us now return to Equation 1.63. Each term in the expansion of a determinant contains exactly one factor from each row and column. All but three columns of the six upper rows of Equation 1.62 contain k; thus, each term in the expansion of Equation 1.62 must contain k to at least the third power. Consequently, Equation 1.62 has at least three zero roots. We shall designate these zero eigen coupling factors by

$$k^{(7)} = k^{(8)} = k^{(9)} = 0 \tag{1.65}$$

If the top six rows of Equation 1.62 and the three right columns are multiplied by -1, the determinant is unchanged, other than that $-k$ is replaced by $+k$. However, this operation has the effect of multiplying the value of the determinant by $(-1)^{6+3} = -1$. Consequently, the polynomial expansion of the determinant may possess only odd powers of k. This means that, after factoring out the three zero solutions, the remaining solutions are of the form

$$\begin{aligned} k^{(1)} &= -k^{(2)} = k_{i1} \\ k^{(3)} &= -k^{(4)} = k_{i2} \\ k^{(5)} &= -k^{(6)} = k_{i3} \end{aligned} \tag{1.66}$$

The k_{i1}, k_{i2}, and k_{i3} used here are the same found in the work of Berlincourt *et al.*[29]

Now let the coupling eigenstate associated with $k^{(A)}$ be designated by $[K_D^{(A)}]$, or simply by $[K^{(A)}]$. In other words, let us have

$$[K^{(A)}] = \begin{bmatrix} T_v^{(A)} \\ E_m^{(A)} \end{bmatrix} \tag{1.67}$$

Thus, according to Equations 1.59 to 1.63, we conclude

$$[M^1][K^{(A)}] = k^{(A)}[M^0][K^{(A)}] \tag{1.68}$$

If $k^{(A)} \neq k^{(B)}$, we can show the usual orthogonality of these eigenvectors (i.e., eigenstates)[32]: Because $[M^1]$ is symmetric, it is true that

$$[K_t^{(A)}][M^1][K^{(B)}] = [K_t^{(B)}][M^1][K^{(A)}] \tag{1.69}$$

However, Equation 1.68 then implies

$$k^{(A)}[K_t^{(A)}][M^0][K^{(B)}] = k^{(B)}[K_t^{(A)}][M^0][K^{(B)}] \tag{1.70}$$

where we have also utilized the symmetry of $[M^0]$. But if $[K^{(A)}]$ and $[K^{(B)}]$ are eigenstates associated with different eigenvalues, Equation 1.70 requires

$$[K_t^{(A)}][M^0][K^{(B)}] = 0 \tag{1.71}$$

Equation 1.68 then, in turn, yields

$$[K_t^{(A)}][M^1][K^{(B)}] = 0 \tag{1.72}$$

Consequently, we see that the coupling eigenstates associated with different coupling eigenfactors are orthogonal with respect to both $[M^0]$ and $[M^1]$.

The eigenstates so far have been expressed only to within a multiplicative factor. It will subsequently be convenient, however, to normalize them so that

$$[K_t^{(A)}][M^0][K^{(A)}] = 1 \tag{1.73}$$

This operation corresponds to requiring the normalized eigenstates to have an elastic plus dielectric energy density of one-half.

Let us now construct the so-called modal matrix having $[K^{(B)}]$ as columns,

$$[\mathscr{K}] = [\mathscr{K}_{AB}] = [K_A^{(B)}] \tag{1.74}$$

(This matrix has several uses that we will develop shortly.) Equations 1.71, 1.73, and 1.74 then yield the result

$$[\mathscr{K}_t][M^0][\mathscr{K}] = [I] \tag{1.75}$$

where $[I]$ is the identity matrix. Alternatively, Equations 1.68, 1.72, 1.73, and 1.74 show

$$[\mathscr{K}_t][M^1][\mathscr{K}] = [k] = [k_{AC}] \tag{1.76}$$

where $[k]$ is a diagonal matrix described by

$$k_{AB} = \begin{cases} 0 & A \neq B \\ k^{(A)} & A = B \end{cases} \tag{1.77}$$

Any arbitrary state vector $[F]$ may be expressed as a linear combination of the nine coupling eigenstates. In matrix algebra, if $[C] = [C_A]$ is the coefficient vector required to express the arbitrary $[F]$ in terms of the eigenstates $[K^{(A)}]$, then $[F]$ and $[C]$ are related by the earlier-defined modal matrix,

$$[F] = [\mathscr{K}][C] \tag{1.78}$$

In view of Equation 1.75, this relation may be solved for $[C]$:

$$[C] = [\mathscr{K}]^{-1}[F] = [\mathscr{K}_t][M^0][F] \tag{1.79}$$

This equation enables one to reduce any state vector to a linear combination of the coupling eigenstates.

In terms of the eigenstates of the material, the energy densities associated with the arbitrary state $[F]$ have a particularly simple form. By virtue of Equations 1.64, 1.75, and 1.78, we have the sum of the elastic and dielectric energies as

$$\begin{aligned} U_E + U_D &= \tfrac{1}{2}[F_t][M^0][F] \\ &= \tfrac{1}{2}[C_t][\mathscr{K}_t][M^0][\mathscr{K}][C] \\ &= \tfrac{1}{2}[C_t][I][C] \\ &= \frac{1}{2}\sum_{A=1}^{9}(C_A)^2 \end{aligned} \tag{1.80}$$

The coupling energy may be similarly expressed, using Equation 1.76 instead of Equation 1.75,

$$\begin{aligned} 2U_M &= \tfrac{1}{2}[F_t][M^1][F] \\ &= \frac{1}{2}\sum_{A=1}^{9} k^{(A)}(C_A)^2 \end{aligned} \tag{1.81}$$

Equation 1.65 illustrates that this last summation may be shortened to run 1 to 6.

The interesting thing to observe in Equations 1.80 and 1.81 is the complete absence of cross terms in the C's when the energy densities are expressed with the coupling eigenstate amplitudes as independent variables.

1.4.2. Example: The Eigen Coupling Factors of a Ferroelectric Ceramic

Let us now demonstrate and carry out a specific example of these manipulations. For a ferroelectric ceramic, according to Table 1.1, class C_{6v}, Equation 1.62 becomes

$$\begin{vmatrix} -ks_{11}^E & -ks_{12}^E & -ks_{13}^E & 0 & 0 & 0 & 0 & 0 & d_{31} \\ -ks_{12}^E & -ks_{11}^E & -ks_{13}^E & 0 & 0 & 0 & 0 & 0 & d_{31} \\ -ks_{13}^E & -ks_{13}^E & -ks_{33}^E & 0 & 0 & 0 & 0 & 0 & d_{33} \\ 0 & 0 & 0 & -ks_{44}^E & 0 & 0 & 0 & d_{15} & 0 \\ 0 & 0 & 0 & 0 & -ks_{44}^E & 0 & d_{15} & 0 & 0 \\ 0 & 0 & 0 & 0 & 0 & -ks_{66}^E & 0 & 0 & 0 \\ 0 & 0 & 0 & 0 & d_{15} & 0 & -k\epsilon_{11}^T & 0 & 0 \\ 0 & 0 & 0 & d_{15} & 0 & 0 & 0 & -k\epsilon_{11}^T & 0 \\ d_{31} & d_{31} & d_{33} & 0 & 0 & 0 & 0 & 0 & -k\epsilon_{33}^T \end{vmatrix} = 0 \tag{1.82}$$

This equation may be expanded in terms of cofactors to obtain

$$ks_{66}^E(k^2\epsilon_{11}^T s_{44}^E - d_{15}^2)^2 \begin{vmatrix} -ks_{11}^E & -ks_{12}^E & -ks_{13}^E & d_{31} \\ -ks_{12}^E & -ks_{11}^E & -ks_{13}^E & d_{31} \\ -ks_{13}^E & -ks_{13}^E & -ks_{33}^E & d_{33} \\ d_{31} & d_{31} & d_{33} & -k\epsilon_{33}^T \end{vmatrix} = 0 \tag{1.83}$$

where the remaining 4×4 determinant corresponds to the 1, 2, 3, 9 elements of the original determinant.

Next we state a theorem of linear algebra that is frequently useful in evaluating the eigenstates once the eigen coupling factors are known.

THEOREM: *Assume some nondegenerate solution $k^{(A)}$ has been found for the eigen coupling factor, Equation 1.62. Then the associated coupling eigenstate may be expressed as*[33]

$$[K_B^{(A)}] = \alpha^{(C,A)}[\Delta_{CB}(k^{(A)})] \tag{1.84}$$

where $\Delta_{CB}(k^{(A)})$ are the signed cofactors of $[M^1 - k^{(A)}M^0]$, and where $\alpha^{(C,A)}$ is a normalization coefficient. The index C appearing in Equation 1.84 may be assigned any value from one to nine, although some values of C may yield trivial solutions for the eigenstates.

One of the zero eigen coupling factors of Equation 1.83, which we shall call $k^{(7)}$, originates with the quantity (ks_{66}^E) occurring as the 6 element of the original determinant. It may be seen by direct substitution into Equation 1.68, using the coefficient matrix of Equation 1.82, that the corresponding eigenstate is

$$[K_t^{(7)}] = \alpha_7[0,0,0,0,0,1,0,0,0] \tag{1.85}$$

This result may also be obtained by application of the theorem, Equation 1.84, with $C = 6$ provided k is not set equal to $k^{(7)} = 0$ until after the normalization coefficient $\alpha^{(6,7)}$ has been evaluated. It is necessary to take limits in this order, because $k^{(7)}$ is a triply degenerate eigen coupling factor; therefore, the theorem is not applicable without modification. The normalization factor α_7, which does not correspond to $\alpha^{(6,7)}$ of Equation 1.84, may be evaluated by means of Equation 1.73 to yield

$$[K_t^{(7)}] = [0,0,0,0,0,(s_{66}^E)^{-1/2},0,0,0] \tag{1.86}$$

From Equation 1.83, four other eigen coupling factors are evident:

$$k^{(1)} = -k^{(2)} = k^{(3)} = -k^{(4)} = \frac{d_{15}}{(\epsilon_{11}^T s_{44}^E)^{1/2}} \tag{1.87}$$

A degeneracy is present here also; thus the eigenstates may not be orthogonal and Equation 1.84 is again not strictly applicable. However, the $k^{(1)}$ and $k^{(2)}$ roots may be considered to originate with the (4,8) elements of the original determinant. Also, $k^{(3)}$ and $k^{(4)}$ may be associated with the (5,7) elements. Consequently we are led to suspect that $[K^{(1)}]$ and $[K^{(2)}]$ have all components zero except the fourth and the eighth, while $[K^{(3)}]$ and $[K^{(4)}]$ have all components zero except the fifth and the seventh. Direct substitution into the form of Equation 1.68 relevant to ferroelectric ceramics confirms this suspicion and leads to the orthogonal eigenstates

$$\begin{aligned}
[K_t^{(1)}] &= [0,0,0,(2s_{44}^E)^{-1/2},0,0,0,(2\epsilon_{11}^T)^{-1/2},0] \\
[K_t^{(2)}] &= [0,0,0,(2s_{44}^E)^{-1/2},0,0,0,-(2\epsilon_{11}^T)^{-1/2},0] \\
[K_t^{(3)}] &= [0,0,0,0,(2s_{44}^E)^{-1/2},0,(2\epsilon_{11}^T)^{-1/2},0,0] \\
[K_t^{(4)}] &= [0,0,0,0,(2s_{44}^E)^{-1/2},0,-(2\epsilon_{11}^T)^{-1/2},0,0]
\end{aligned} \tag{1.88}$$

The remaining 4×4 determinant has two zero roots and two nonzero roots. Its expansion is

$$k^4\epsilon_{33}^T[s_{33}^E(s_{11}^E + s_{12}^E) - 2(s_{13}^E)^2] + k^2[4d_{31}d_{33}s_{13}^E - 2d_{31}^2s_{33}^E - d_{33}^2(s_{11}^E + s_{12}^E)] = 0 \tag{1.89}$$

Consequently, we find

$$k^{(8)} = k^{(9)} = 0$$

$$k^{(5)} = -k^{(6)} = \left[\frac{2d_{31}^2 s_{33}^E - 4d_{31}d_{33}s_{13}^E + d_{33}^2(s_{11}^E + s_{12}^E)}{\epsilon_{33}^T[s_{33}^E(s_{11}^E + s_{12}^E) - 2(s_{13}^E)^2]}\right]^{1/2} \tag{1.90}$$

The second of these relations may alternatively be expressed as

$$k^{(5)} = -k^{(6)} = \left[\frac{\epsilon_{33}^T - \epsilon_{33}^S}{\epsilon_{33}^T}\right]^{1/2} \tag{1.91}$$

The eigenstates associated with $k^{(8)}$ and $k^{(9)}$ are not necessarily orthogonal, because a degeneracy is also present here. As $k^{(8)}$ and $k^{(9)}$ are generated by the (1,2,3,9) elements of the original coefficient matrix, it will be true that these eigenstates are solutions of

$$\begin{bmatrix} 0 & 0 & 0 & d_{31} \\ 0 & 0 & 0 & d_{31} \\ 0 & 0 & 0 & d_{33} \\ d_{31} & d_{31} & d_{33} & 0 \end{bmatrix} \begin{bmatrix} K_1^{(8,9)} \\ K_2^{(8,9)} \\ K_3^{(8,9)} \\ K_9^{(8,9)} \end{bmatrix} = [0] \tag{1.92}$$

with

$$K_A^{(8)} = K_A^{(9)} = 0 \qquad A \neq 1, 2, 3, 9 \tag{1.93}$$

Equation 1.92 in turn requires

$$K_9^{(8)} = K_9^{(9)} = 0 \tag{1.94}$$

and

$$d_{31}K_1^{(8,9)} + d_{31}K_2^{(8,9)} + d_{33}K_3^{(8,9)} = 0 \tag{1.95}$$

Let us specify the $[K^{(8)}]$ solution to Equations 1.94 and 1.95 to be of the form

$$[K_t^{(8)}] = \alpha_8[1,-1,0,0,0,0,0,0,0] \tag{1.96}$$

Normalizing this eigenstate, we have

$$[K_t^{(8)}] = [(s_{66}^E)^{-1/2},-(s_{66}^E)^{-1/2},0,0,0,0,0,0,0] \tag{1.97}$$

where use is made of the relation given in Table 1.1,

$$s_{66}^E = 2(s_{11}^E - s_{12}^E) \tag{1.98}$$

Consequently, it may be observed that $[K^{(8)}]$ is just $[K^{(7)}]$ rotated by -45 degrees in the x_1x_2-plane. The eigenstates $[K^{(7)}]$ and $[K^{(8)}]$ represent pure shear in the x_1x_2-plane. As the coefficient matrix of Equation 1.82 indicates that this type of stress is piezoelectrically decoupled, it is fairly obvious that it should be associated with uncoupled ($k^{(A)} = 0$) eigenstates.

A second normalized solution to Equation 1.95 orthogonal to $[K^{(8)}]$, is

$$[K_t^{(9)}] = \alpha_9\left[\tfrac{1}{2}, \tfrac{1}{2}, -\frac{d_{31}}{d_{33}}, 0, 0, 0, 0, 0, 0\right] \tag{1.99}$$

where

$$\alpha_9 = \left[\tfrac{1}{2}(s_{11}^E + s_{12}^E) - \frac{2d_{31}s_{13}^E}{d_{33}} + \frac{d_{31}^2 s_{33}^E}{d_{33}^2}\right]^{-1/2} \tag{1.100}$$

In the case of this decoupled eigenstate, the dilation in the x_1x_2-plane reacts piezoelectrically with a sign opposite to that of the dilation in the x_3-direction. Magnitudes of the two effects are adjusted by the theory to be equal, so that a total cancellation occurs in U_M.

The two remaining eigenstates $[K^{(5)}]$ and $[K^{(6)}]$ are solutions of

$$\begin{bmatrix} \pm k^{(5)}s_{11}^E & \pm k^{(5)}s_{12}^E & \pm k^{(5)}s_{13}^E & -d_{31} \\ \pm k^{(5)}s_{12}^E & \pm k^{(5)}s_{11}^E & \pm k^{(5)}s_{13}^E & -d_{31} \\ \pm k^{(5)}s_{13}^E & \pm k^{(5)}s_{13}^E & \pm k^{(5)}s_{33}^E & -d_{33} \\ -d_{31} & -d_{31} & -d_{33} & \pm k^{(5)}\epsilon_{33}^T \end{bmatrix} \begin{bmatrix} K_1^{(5,6)} \\ K_2^{(5,6)} \\ K_3^{(5,6)} \\ K_9^{(5,6)} \end{bmatrix} = [0] \tag{1.101}$$

with

$$K_A^{(5)} = K_A^{(6)} = 0 \qquad A \neq 1, 2, 3, 9 \tag{1.102}$$

Equation 1.101 is cumbersome to evaluate for $K_A^{(5,6)}$, although Equation 1.84 is of considerable help. A transformation of independent variables facilitates this operation, however:

$$\begin{bmatrix} K_1^{(5,6)} \\ K_2^{(5,6)} \\ K_3^{(5,6)} \\ K_9^{(5,6)} \end{bmatrix} = \begin{bmatrix} c_{11}^E & c_{12}^E & c_{13}^E & -e_{31} \\ c_{12}^E & c_{11}^E & c_{13}^E & -e_{31} \\ c_{13}^E & c_{13}^E & c_{33}^E & -e_{33} \\ 0 & 0 & 0 & 1 \end{bmatrix} \begin{bmatrix} S_1^{(5,6)} \\ S_2^{(5,6)} \\ S_3^{(5,6)} \\ E_3^{(5,6)} \end{bmatrix} \tag{1.103}$$

where $S_A^{(B)}$ is the strain associated with eigenstate B.

Substitution of Equation 1.103 into Equation 1.101 yields

$$\begin{bmatrix} \pm k^{(5)} & 0 & 0 & -d_{31}(1 \pm k^{(5)}) \\ 0 & \pm k^{(5)} & 0 & -d_{31}(1 \pm k^{(5)}) \\ 0 & 0 & \pm k^{(5)} & -d_{33}(1 \pm k^{(5)}) \\ -e_{31} & -e_{31} & -e_{33} & \pm k^{(5)}\epsilon_{33}^{T}(1 \pm k^{(5)}) \end{bmatrix} \begin{bmatrix} S_1^{(5,6)} \\ S_2^{(5,6)} \\ S_3^{(5,6)} \\ E_3^{(5,6)} \end{bmatrix} = [0] \tag{1.104}$$

Application of the theorem, Equation 1.84, to this result then shows

$$\begin{aligned} S_1^{(5,6)} &= S_2^{(5,6)} = d_{31}(1 \pm k^{(5)}) \\ S_3^{(5,6)} &= d_{33}(1 \pm k^{(5)}) \\ E_3^{(5,6)} &= \pm k^{(5)} \end{aligned} \tag{1.105}$$

This result can be transformed back into $(T_v^{(5,6)}, E_3^{(5,6)})$ by means of Equation 1.103, and the result can be normalized to complete the evaluation of the coupling eigenstates for a ferroelectric ceramic. As this is a singularly unenlightening computation, it is not carried out here.

1.4.3. Application of the Concept of Eigen Coupling to Filter Design. Development of an Unresolved Indeterminacy.

Let us now consider the application of eigen coupling to the design of maximum bandwidth filters which we mentioned at the beginning of Section 1.4.

Piezoelectric filters utilize the acoustic resonances of a piezoelectric element to obtain a sharp frequency response. A typical element is sketched in Figure 1.1, along with its equivalent circuit at frequencies near the acoustic resonance of interest. (The development of this equivalent circuit is discussed in nearly any reference on piezoelectric resonators, as well as in the later chapters of this monograph.)

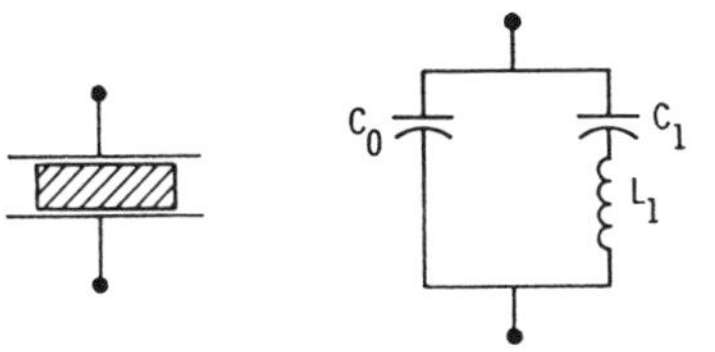

Figure 1.1 A piezelectric resonator and its equivalent circuit in the vicinity of an acoustic resonance.

The bandwidth of a lattice filter that can be constructed with piezoelectric resonators forming its branches is almost directly proportional to the ratio C_1/C_0 of the two equivalent capacitances shown in Figure 1.1.[34] Consequently, the problem of maximizing filter bandwidth is equivalent to the problem of maximizing this ratio.

If we assume we have a piezoelectric element excited at some resonance, we may find the total elastic, electric, and piezoelectric energies of the element at that resonance by integrating U_E, U_D, and U_M over the element,

$$W_{E,D,M} = \iiint U_{E,D,M}\, dV \tag{1.106}$$

A dynamic coupling factor may then be associated with this resonance of the overall element by the definition

$$k_m'^2 = \frac{W_M^2}{W_E W_D} \tag{1.107}$$

Note that the geometric mean of the energies is used here. Lewis[35] has shown that if $k_m'^2$ is "much" smaller than unity, the above all-important capacitance ratio is none other than this $k_m'^2$,

$$\begin{aligned} \frac{C_1}{C_0} &= k_m'^2 + O(k_m'^4) \\ &= \frac{\left(\iiint E_m d_{m\nu} T_\nu\, dV\right)^2}{\left(\iiint E_m \epsilon_{mn}^T E_n\, dV\right)\left(\iiint T_\mu s_{\mu\nu}^E T_\nu\, dV\right)} + O(k_m'^4) \end{aligned} \tag{1.108}$$

Here $O(k_m'^4)$ represents correction terms on the order of $k_m'^4$.

In addition, according to the Schwartz inequality,[36] the integral quotient of Equation 1.108 is bounded by the upper limit

$$\frac{C_1}{C_0} \leq \frac{1}{V} \iiint \left(\frac{U_M^2}{U_D U_E}\right) dV \leq \left(\frac{U_M^2}{U_D U_E}\right)_{\max} \tag{1.109}$$

where the equality holds only if E_m and T_ν are uniform. However, the rightmost term of Equation 1.109 is itself bounded by the largest of the eigen coupling factors. Consequently, for materials in which the largest $k^{(A)2}$ is much smaller than unity, this theory provides a technique for computing the maximum bandwidth that may be attained with a piezoelectric resonator of that material. This maximum is specified by

$$\left(\frac{C_1}{C_0}\right)_{\max} = (k^{(A)^2})_{\max} \tag{1.110}$$

The eigenstate (T'_ν, E'_m) associated with the geometric mean definition of this maximum eigen coupling factor also has a special significance: Any resonator sample shape and electrode configuration that can support a resonant mode characterized by a stress and field distribution coinciding on a point-by-point basis with this eigenstate is optimum with respect to bandwidth maximization. However, the problem of finding such an ideal sample shape and electrode configuration will, in general, have no exact solution. We may thus alternatively say that the problem of bandwidth maximization is reduced to the problem of finding sample shapes and electrode configurations that can support modes deviating (in a mean energy density sense) as little as possible from the ideal distribution.

In cases where $(k^{(A)^2})_{\max}$ is not negligible compared to unity (such as some ferroelectric ceramics, where it may exceed 0.5), the remarks of the last paragraph are not applicable, and certain indeterminancies appear to be present.

Let us explore this situation by inquiring what happens if the independent variables of Equation 1.53, T_ν and E_m, are replaced by S_ν and D_m. Then the energy densities become, according to Equations 1.16 and 1.25,

$$\begin{aligned} 2U &= D_m\beta^S_{mn}D_n - 2D_mh_{m\nu}S_\nu + S_\mu c^D_{\mu\nu}S_\nu \\ &= 2(U'_D - 2U'_M + U'_E) \end{aligned} \tag{1.111}$$

If we now define the algebraic mean coupling factor by

$$k = \frac{2U'_M}{U'_D + U'_E} \tag{1.112}$$

where $U'_{M,D,E}$ are as given in Equation 1.111, the question arises, "Does this definition lead to the same eigen coupling factors and eigenstates as Equation 1.55?"

This question has a rather disturbing answer. The eigen coupling factors are always unchanged, but the eigenstates are modified if the $k^{(A)^2}$ are not negligible compared to unity. In particular, based on the identities given in Section 1.1, one can easily verify that

$$\begin{bmatrix} c^E_{\mu\lambda} & 0 \\ 0 & \beta^T_{nm} \end{bmatrix} \begin{bmatrix} ks^E_{\lambda\nu} & d_{\lambda l} \\ d_{n\nu} & k\epsilon^T_{ml} \end{bmatrix} = \begin{bmatrix} kc^D_{\mu\lambda} & h_{\mu m} \\ h_{n\lambda} & k\beta^S_{nm} \end{bmatrix} \begin{bmatrix} s^D_{\lambda\nu} & 0 \\ 0 & \epsilon^S_{ml} \end{bmatrix} \tag{1.113}$$

Thus, if k is a solution of Equation 1.58, it is also an eigen coupling factor of Equations 1.111 and 1.112, as may readily be seen by taking the determinant of both sides of Equation 1.113 and recalling that $\det([A][B]) = \det[A]\cdot\det[B]$.[37]

However, if $(T_\lambda, E_l)_0$ is an eigenstate of Equation 1.55, Equation 1.113 indicates that

$$\begin{bmatrix} S_v \\ D_m \end{bmatrix}_1 = \begin{bmatrix} s^D_{\lambda v} & 0 \\ 0 & \epsilon^S_{ml} \end{bmatrix} \begin{bmatrix} T_\lambda \\ E_l \end{bmatrix}_0 \tag{1.114}$$

is the corresponding eigenstate of Equation 1.112. If this is reversed, we see

$$\begin{bmatrix} T_\lambda \\ E_l \end{bmatrix}_0 = \begin{bmatrix} c^D_{\lambda v} & 0 \\ 0 & \beta^S_{ml} \end{bmatrix} \begin{bmatrix} S_v \\ D_m \end{bmatrix}_1 \tag{1.115}$$

This, however, is a disturbing result, because the (T_λ, E_l) representation of eigenstate $(S_v, D_m)_1$ is actually, according to Equation 1.25,

$$\begin{bmatrix} T_\lambda \\ E_l \end{bmatrix}_1 = \begin{bmatrix} c^D_{\lambda v} & -h_{\lambda m} \\ -h_{lv} & \beta^S_{ml} \end{bmatrix} \begin{bmatrix} S_v \\ D_m \end{bmatrix}_1 \tag{1.116}$$

In other words, the coupling eigenstates of a material depend on the choice of independent variables unless the piezoelectric effects, given by h in Equation 1.116, are quite small.

In the case of $(k^{(A)^2})_{\max}$ not small compared to unity, it is more correct to say that maximum filter bandwidth is proportional to C_1/C_T (where C_T is the dc capacitance of the piezoelectric element) rather than to C_1/C_0. Consequently, it then becomes desirable to maximize this ratio, which is apparently bounded by

$$(C_1/C_T)_{\max} \leq (k^{(A)^2})_{\max} \tag{1.117}$$

While this new inequality is almost certainly valid, it has apparently never been verified for the present situation of large $k^{(A)^2}$.

We can thus still get an idea of how wide a bandwidth can be achieved with a filter constructed of strongly piezoelectric resonators, as $(k^{(A)^2})_{\max}$ does not depend on the choice of independent variables. However, the present-day theory apparently does not yield satisfactory information on what stress-field distribution to use to attain this optimum result for the strong coupling case. This is true because of the inconsistency between the stress-field distributions of the optimum eigenstate as predicted by Equation 1.115 on one hand and Equation 1.116 on the other. There is clearly a need for penetrating work on this problem.

Part of this indeterminacy doubtless revolves around the following difficulty: While Lewis[35] has shown that for $(k^{(A)^2})_{\max}$ small,

$$\frac{C_1}{C_0} \doteq \frac{\left(\iiint E_m d_{m\nu} T_\nu \, dV\right)^2}{\left(\iiint E_m \epsilon^T_{mn} E_n \, dV\right)\left(\iiint T_\mu s^E_{\mu\nu} T_\nu \, dV\right)}$$

$$\doteq \frac{C_1}{C_T} \doteq \frac{\left(\iiint D_m h_{m\nu} S_\nu \, dV\right)^2}{\left(\iiint D_m \beta^S_{mn} D_n \, dV\right)\left(\iiint S_\mu c^D_{\mu\nu} S_\nu \, dV\right)} \tag{1.118}$$

none of these four quantities are nearly equal for large coupling. Consequently, if the Schwartz inequality were applied to maximize the two integral quotients in Equation 1.118, the two results would describe different states of the material.

A still more basic but not widely known fact, and probable cause of the trouble, is that if we define k' (or k) by Equations 1.53 and 1.111, the two resulting values will differ by $O(k^4)$ — they are not equal as is frequently believed. For example, let us take a typical coefficient matrix with relatively strong piezoelectric coupling:

$$[M] = \left[\begin{array}{cccccc|ccc}
1.0 & -0.3 & -0.3 & 0.1 & 0.1 & 0.1 & 0.0 & 0.2 & 0.1 \\
-0.3 & 1.0 & -0.3 & 0.1 & 0.1 & -0.1 & 0.2 & 0.1 & 0.1 \\
-0.3 & -0.3 & 1.0 & 0.1 & 0.1 & 0.1 & 0.1 & 0.1 & 0.3 \\
0.1 & 0.1 & 0.1 & 0.6 & -0.1 & 0.1 & 0.2 & 0.3 & 0.2 \\
0.1 & 0.1 & 0.1 & -0.1 & 0.4 & 0.1 & 0.3 & 0.2 & 0.1 \\
0.1 & -0.1 & 0.1 & 0.1 & 0.1 & 0.5 & 0.4 & 0.0 & 0.1 \\
\hline
0.0 & 0.2 & 0.1 & 0.2 & 0.3 & 0.4 & 1.0 & 0.3 & 0.2 \\
0.2 & 0.1 & 0.1 & 0.3 & 0.2 & 0.0 & 0.3 & 1.0 & 0.1 \\
0.1 & 0.1 & 0.3 & 0.2 & 0.1 & 0.1 & 0.2 & 0.1 & 0.8
\end{array}\right] \tag{1.119}$$

and the 9 typical state vectors,

$$[F_t]_1 = [1,2,3,4,5,6,7,8,9]$$
$$[F_t]_2 = [2,3,4,5,6,7,8,9,1]$$
$$\vdots$$
$$[F_t]_9 = [9,1,2,3,4,5,6,7,8] \tag{1.120}$$

Utilizing a digital computer, we may then obtain the values given in Table 1.7 for k and k' according to Equations 1.53 and 1.111 for each of these 9 states. The eigen coupling factors for this matrix are not negligible compared to unity: ± 0.7731, ± 0.6088, ± 0.3814. We can see from the table of results that the coupling factors associated with a given state came out quite different.

On the other hand, if the 3×6 and 6×3 piezoelectric submatrices of Equation 1.119 are all reduced by a factor of 100 while the other two submatrices are unchanged, the eigen coupling factors become indeed much less than unity: ± 0.007731, ± 0.006088, ± 0.003814. If the coupling factors associated with the nine state vectors of Equation 1.120 are computed for this new coefficient matrix, the values result as given in Table 1.8. Here we observe that k and k' are relatively insensitive to change in definition of k and k' from Equations 1.53 to 1.111. This insensitivity is to be expected for small k, based on the earlier claim that the discrepancy is $O(k^4)$.

Table 1.7. Electroelastic Coupling Factors for $[M]$ as Given by Equation 1.119 with Respect to the $[F]$ State Vectors of Equation 1.120

	k(Eq. 1.53)	k(Eq. 1.111)	k'(Eq. 1.53)	k'(Eq. 1.111)
$[F]_1$	0.5626	0.7595	0.6959	0.7626
$[F]_2$	0.6175	0.7612	0.6510	0.7620
$[F]_3$	0.6965	0.7667	0.7098	0.7670
$[F]_4$	0.3518	0.7489	0.6724	0.7585
$[F]_5$	0.5493	0.7444	0.7279	0.7488
$[F]_6$	0.6019	0.7305	0.6698	0.7325
$[F]_7$	0.6506	0.7390	0.6513	0.7391
$[F]_8$	0.5111	0.7255	0.5118	0.7255
$[F]_9$	0.4846	0.7259	0.4960	0.7265

Table 1.8. Electroelastic Coupling Factors for $[M]$ as Given by Equation 1.119 with Piezoelectric Coefficients Reduced by a Factor of 100

	k(Eq. 1.53)	k(Eq. 1.111)	k'(Eq. 1.53)	k'(Eq. 1.111)
$[F]_1$	0.005626	0.005675	0.006960	0.006979
$[F]_2$	0.006175	0.006207	0.006510	0.006535
$[F]_3$	0.006965	0.006981	0.007098	0.007111
$[F]_4$	0.003519	0.003604	0.006724	0.006761
$[F]_5$	0.005493	0.005504	0.007279	0.007280
$[F]_6$	0.006019	0.006043	0.006698	0.006705
$[F]_7$	0.006506	0.006521	0.006513	0.006527
$[F]_8$	0.005111	0.005144	0.005118	0.005150
$[F]_9$	0.004846	0.004887	0.004961	0.005000

1.4.4. Eigen Loss Tangents

For a piezoelectric material excited linearly in sinusoidal steady state, the most general phenomenological loss is represented by making all the material coefficients complex, as in Equation 1.45:

$$[M] = [M'] - j[M''] = \begin{bmatrix} s_{\mu\nu}^{E} & d_{\mu n} \\ d_{m\nu} & \epsilon_{mn}^{T} \end{bmatrix} \tag{1.121}$$

The components of $[M'']$ are much smaller than the corresponding components of $[M']$.

As in the first parts of Section 1.4, let us denote the stress-field state vectors by $[F]$, where a suppressed $e^{j\omega t}$ dependence is now implied. Let us denote the complex conjugate of $[F]$ by $[F^*]$, and the Hermitian conjugate of $[F]$ by

$$[F^H] = [F_t^*] \tag{1.122}$$

The time-average electroelastic energy storage density may now be represented as

$$\begin{aligned} U &= \tfrac{1}{4}\,\mathrm{Re}\,\{[F^H][M][F]\} \\ &= \tfrac{1}{4}[F^H][M'][F] \end{aligned} \tag{1.123}$$

Power dissipation density is similarly described, according to Equation 1.43, as

$$\begin{aligned} P_d &= \tfrac{1}{2}\omega\,\mathrm{Im}\,\{[F^H][M][F]\} \\ &= \tfrac{1}{2}\omega[F^H][M''][F] \end{aligned} \tag{1.124}$$

A loss tangent δ is associated with each state vector $[F]$ by the definition

$$P_d = 2\omega U\delta \tag{1.125}$$

(Note that if traveling wave or resonance conditions prevail, U is only half the total energy; the other half is kinetic. This factor of $\frac{1}{2}$ is compensated by the factor of 2 in Equation 1.125 so that Equation 1.125 is actually equivalent to conventional definitions of loss tangents.[38,39])

It is possible to define loss eigenstates $[L^{(A)}]$ which are quite analogous to the coupling eigenstates of the first parts of this section. When U and P_d are described in terms of the loss eigenstate amplitudes, expressions analogous to Equations 1.80 and 1.81 are obtained in which cross terms are absent.

Let us specify the loss eigenstates $[L^{(A)}]$ and eigen loss tangents $\delta^{(A)}$ by an equation analogous to Equation 1.62,

$$[M'' - \delta^{(A)}M'][L^{(A)}] = [0] \tag{1.126}$$

Unlike $[M^1]$, $[M']$ and $[M'']$ are both positive definite. Consequently, unlike $k^{(A)}$, the $\delta^{(A)}$ will all be positive and real.

Considerations analogous to Equations 1.68 to 1.72 indicate that loss eigenstates associated with different eigen loss tangents are orthogonal with respect to both $[M']$ and $[M'']$:

$$\begin{aligned} [L^{(A)H}][M'][L^{(B)}] &= 0 \\ [L^{(A)H}][M''][L^{(B)}] &= 0 \end{aligned} \tag{1.127}$$

The analogy between eigen loss quantities and eigen coupling quantities may be extended by normalizing the $[L^{(A)}]$ according to

$$[L^{(A)H}][M'][L^{(A)}] = 1 \tag{1.128}$$

The $[L^{(A)}]$ may be defined to have all real components. In particular, assume some $[L^{(A)}]$ with complex components satisfies Equation 1.126. Then, $[L^{(A)}*]$ also satisfies Equation 1.126 with the same $\delta^{(A)}$, as $\delta^{(A)}$ must be real. Consequently, either $\delta^{(A)}$ is a degenerate eigen loss tangent, or $[L^{(A)}]$ and $[L^{(A)}*]$ differ by a multiplicative constant only. In the second case, that constant may be chosen to have zero phase, thus making $[L^{(A)}]$ pure real without violating Equation 1.128. In the first case, the two complex degenerate loss eigenstates may each be made pure real by the transformation

$$\begin{aligned} [L^{(A_1)}] &= \tfrac{1}{2}\,\mathrm{Re}\,\{[L^{(A)}] + [L^{(A)}*]\} \\ [L^{(A_2)}] &= \tfrac{1}{2}\,\mathrm{Im}\,\{[L^{(A)}] - [L^{(A)}*]\} \end{aligned} \tag{1.129}$$

A so-called loss modal matrix $[\mathscr{L}]$ may be constructed with $[L^{(B)}]$ as columns

$$[\mathscr{L}] = [\mathscr{L}_{AB}] = [L_A^{(B)}] \tag{1.130}$$

This matrix has several quite useful properties which will soon be apparent. Its nature is similar to $[\mathscr{K}]$ as defined in the first part of this section:

$$[\mathscr{L}_t][M'][\mathscr{L}] = [I] \tag{1.131}$$

and

$$[\mathscr{L}_t][M''][\mathscr{L}] = [\delta] = [\delta_{AB}] \tag{1.132}$$

where

$$\delta_{AB} = \begin{cases} 0 & A \neq B \\ \delta^{(A)} & A = B \end{cases} \tag{1.133}$$

Note that δ_{AB} is not the Kronecker delta here.

An arbitrary state vector may be expressed as a linear combination of the nine loss eigenstates. In matrix algebra, if $[B] = [B_A]$ is the coefficient vector required to express a given $[F]$ in terms of the eigenstates $[L^{(A)}]$, then

$$[F] = [\mathscr{L}][B] \tag{1.134}$$

Unlike the $[C]$ coefficient vector associated with the coupling eigenstates in the dc case, this $[B]$ coefficient vector may be complex. In view of Equation 1.131, Equation 1.134 may be solved for $[B]$:

$$[B] = [\mathscr{L}]^{-1}[F] = [\mathscr{L}_t][M'][F] \tag{1.135}$$

By means of Equations 1.135 any state vector may be expressed as a linear combination of the loss eigenstates.

As we stated before, the energy and power dissipation densities may be expressed as the sum of the squares of the loss eigenstate amplitudes. Equations 1.123, 1.131, and 1.134 indicate that

$$\begin{aligned} U &= \tfrac{1}{4}[F^H][M'][F] \\ &= \tfrac{1}{4}[B^H][\mathscr{L}_t][M'][\mathscr{L}][B] \\ &= \tfrac{1}{4}[B^H][I][B] \\ &= \tfrac{1}{4}\sum_{A=1}^{9} |B_A|^2 \end{aligned} \tag{1.136}$$

The power dissipation density counterpart of Equation 1.136 is obtained similarly, using Equation 1.124 instead of Equation 1.123 and Equation 1.132 instead of Equation 1.131:

$$\begin{aligned} P_d &= \tfrac{1}{2}\omega[F^H][M''][F] \\ &= \tfrac{1}{2}\omega \sum_{A=1}^{9} \delta^{(A)} \, |B_A|^2 \end{aligned} \tag{1.137}$$

It is interesting to observe the similarity between Equations 1.136 and 1.80 for $U_E + U_D$. Likewise, Equation 1.137 for P_d is analogous to Equation 1.81 for U_M. All four summations are characterized by a complete absence of cross terms when state vectors are represented as linear combinations of the appropriate eigenstates.

It is now possible to demonstrate that the eigen loss tangents have stationary properties that are quite similar to the stationary properties of the eigen coupling factors. In particular, assume that we have a state vector which is a small perturbation of a loss eigenstate:

$$[F] = [L^{(A)}] + \sum_{B=1}^{9} \gamma_B [L^{(B)}] \qquad \gamma_B \ll 1 \tag{1.138}$$

Substitution of the eigenstate coefficients of Equation 1.138 into δ as defined by Equations 1.125, 1.136, and 1.137 yields

$$\delta = \frac{P_d}{2\omega U} = \frac{(1 + \gamma_A)^2 \delta^{(A)} + \sum_{B \neq A} \gamma_B^2 \delta^{(B)}}{(1 + \gamma_A)^2 + \sum_{B \neq A} \gamma_B^2} \tag{1.139}$$

Consequently, we see that a first-order variation in $[F]$ around some loss eigenstate produces only a second-order variation in δ. This result is identical to Equation 1.59 defining the coupling eigenstates and eigenfactors; there a first-order variation in $[F]$ around some coupling eigenstate produces only a second-order variation in k.

1.4.5. Example: The Eigen Loss Tangents of the Ferroelectric Ceramic, PZT 65/35

There is apparently no piezoelectric material for which all the elements of $[M'']$ are known. However, for the hot-pressed lead-zirconate titanate ceramic described in Section 1.3, PZT 65/35, the (1,2,6,9) submatrix of $[M'']$ has been measured as a function of relative polarization.[27] (Refer to the work of Land, *et al.*,[23] for a definition of relative polarization and for information on the exact composition of PZT 65/35.[4])

Consequently, for the purpose of illustration, we shall now discuss the eigen loss tangents and loss eigenstates of PZT 65/35 with $[F]$ constrained to be of the form

$$[F_t] = [T_1,T_2,0,0,0,T_6,0,0,E_3] \tag{1.140}$$

It should be pointed out that the four eigenstates and loss tangents obtained in this manner will not generally correspond to any of the nine eigenstates and loss tangents obtained with unconstrained $[F]$.

The version of Equation 1.126 relevant for this computation is

$$\begin{bmatrix} s_{11}^{E''} - \delta s_{11}^{E'} & s_{12}^{E''} - \delta s_{12}^{E'} & 0 & d_{31}'' - \delta d_{31}' \\ s_{12}^{E''} - \delta s_{12}^{E'} & s_{11}^{E''} - \delta s_{11}^{E'} & 0 & d_{31}'' - \delta d_{31}' \\ 0 & 0 & s_{66}^{E''} - \delta s_{66}^{E'} & 0 \\ d_{31}'' - \delta d_{31}' & d_{31}'' - \delta d_{31}' & 0 & \epsilon_{33}^{T''} - \delta \epsilon_{33}^{T'} \end{bmatrix} \begin{bmatrix} T_1 \\ T_2 \\ T_6 \\ E_3 \end{bmatrix} = [0] \tag{1.141}$$

One eigen loss tangent is immediately obvious from Equation 1.141:

$$\delta^{(7)} = s_{66}^{E''}/s_{66}^{E'} \tag{1.142}$$

The index "7" is chosen in analogy with the corresponding eigen coupling factor described by Equation 1.86. The loss eigenstate associated with $\delta^{(7)}$, normalized in accordance with Equation 1.128, is

$$[L_t^{(7)}] = [0,0,0,0,0,(s_{66}^{E'})^{-1/2},0,0,0] \tag{1.143}$$

Direct substitution into Equation 1.141 indicates that

$$\delta^{(8)} = s_{66}^{E''}/s_{66}^{E'} = (s_{11}^{E''} - s_{12}^{E''})/(s_{11}^{E'} - s_{12}^{E'}) \tag{1.144}$$

$$[L_t^{(8)}] = [(s_{66}^{E'})^{-1/2},-(s_{66}^{E'})^{-1/2},0,0,0,0,0,0,0] \tag{1.145}$$

constitute a second solution. This loss eigenstate corresponds to the coupling eigenstate $[K^{(8)}]$ as described by Equation 1.97. It may be seen to be just $[L^{(7)}]$ rotated by -45 degrees in the x_1x_2-plane.

Expansion of the determinant associated with Equation 1.141 indicates that the two remaining eigen loss tangents are

$$\delta^{(1,2)} = \frac{(\delta_+ + \delta_\epsilon - 2k_p^2\delta_d)}{2(1 - k_p^2)} \pm \frac{[(\delta_+ + \delta_\epsilon - 2k_p^2\delta_d)^2 - 4(\delta_+\delta_\epsilon - k_p^2\delta_d^2)(1 - k_p^2)]^{1/2}}{2(1 - k_p^2)} \tag{1.146}$$

where the following definitions have been made:

$$\delta_+ = (s_{11}^{E''} + s_{12}^{E''})/(s_{11}^{E'} + s_{12}^{E'}) \tag{1.147}$$

$$\delta_\epsilon = \epsilon_{33}^{T''}/\epsilon_{33}^{T'} \tag{1.148}$$

$$\delta_d = d_{31}''/d_{31}' \tag{1.149}$$

and

$$k_p^2 = \frac{2d_{31}'^2}{(s_{11}^{E'} + s_{12}^{E'})\epsilon_{33}^{T'}} \tag{1.150}$$

For PZT 65/35, s_{11}^E and s_{12}^E have been found to have the same phase angle.[27] By virtue of Equation 1.98, s_{66}^E also must have the same phase angle, and this angle will be δ_+ as defined by Equation 1.147. In Figure 1.2, we have plotted δ_+, δ_ϵ, δ_d, and k_p^2 versus relative polarization for

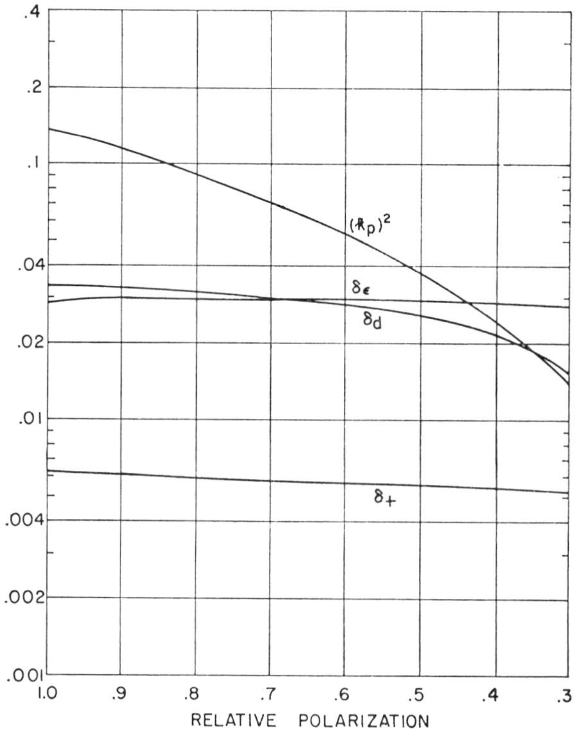

Figure 1.2. The planar coupling factor $(k_p)^2$ and the loss tangents (δ_ϵ = dielectric loss tangent, δ_d = piezoelectric loss tangent, δ_+ = elastic loss tangent) for PZT 65/35 as a function of polarization state.

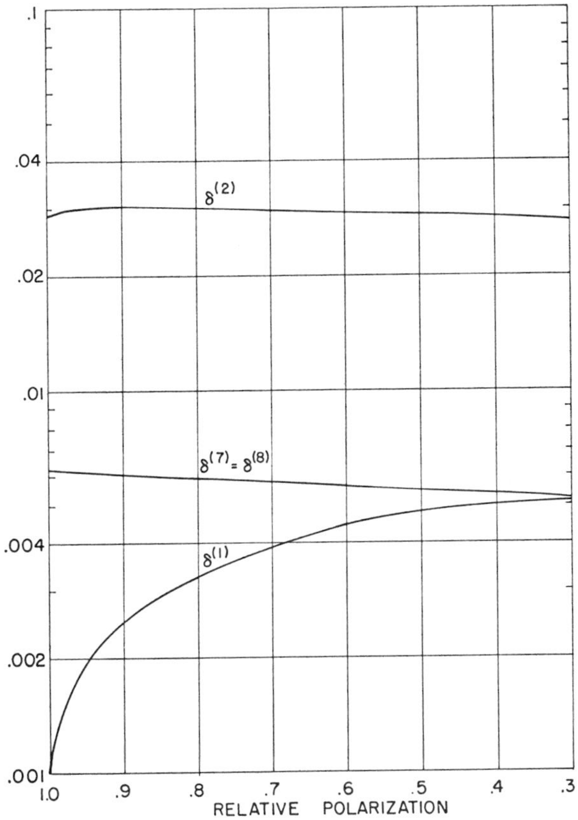

Figure 1.3. The eigen loss tangents for plane-stressed PZT 65/35 as a function of polarization state.

PZT 65/35. The result of substituting these values into Equations 1.142, 1.144, and 1.146 yields the eigen loss tangents $\delta^{(1)}$, $\delta^{(2)}$, $\delta^{(7)}$, and $\delta^{(8)}$ illustrated in Figure 1.3 as a function of relative polarization. The various energy density components associated with the corresponding normalized loss eigenstates are shown in Figure 1.4 for $\mathscr{L}^{(1)}$ and $\mathscr{L}^{(2)}$; $\mathscr{L}^{(7)}$ and $\mathscr{L}^{(8)}$ are purely elastic shear, and consequently have

$$\begin{aligned} U_E^{(7)} &= U_E^{(8)} = \tfrac{1}{4} \\ U_D^{(7)} &= U_D^{(8)} = 0 \end{aligned} \tag{1.151}$$

(The quantities $U_D^{(A)}$ and $U_E^{(A)}$ are defined in analogy with Equations

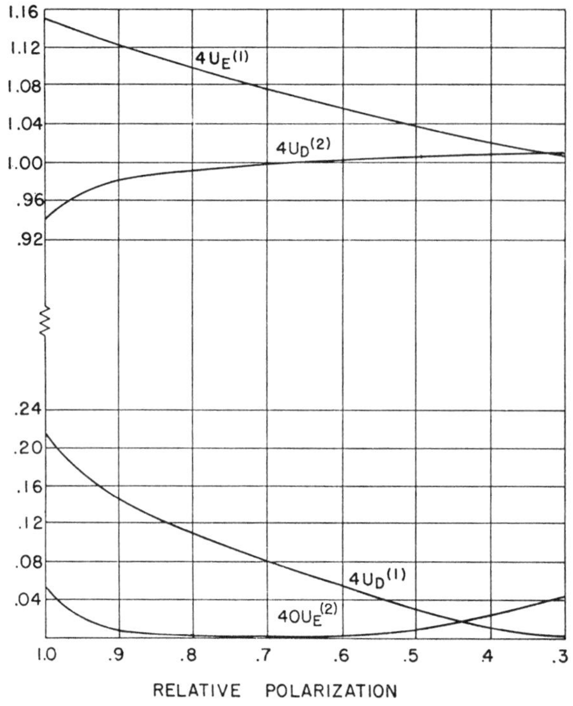

Figure 1.4. The energy density components associated with the plane-stress loss eigenstates of PZT 65/35. (Here $U_D^{(A)}$ is the dielectric energy density and $U_E^{(A)}$ is the elastic energy density. Not shown is the piezoelectric energy density, $U_M^{(A)} = \frac{1}{8} - \frac{1}{2}U_D^{(A)} - \frac{1}{2}U_M^{(A)}$.)

1.53.) It should be observed that Equations 1.123 and 1.128 require

$$U_D^{(A)} + 2U_M^{(A)} + U_E^{(A)} = \tfrac{1}{4} \tag{1.152}$$

so that once $U_D^{(A)}$ and $U_E^{(A)}$ are given, as in Figure 1.4, $U_M^{(A)}$ is automatically determined.

Perhaps the most interesting observation to be made from these figures is that the eigen loss tangents may be considerably smaller than any of the loss tangents associated with the individual elements of [M]. This effect is best observed in the behavior of $\delta^{(1)}$ near unity relative polarization (Figure 1.3).

1.4.6. Application of the Concept of Eigen Loss Tangents to Filter Design

As we discussed in Section 1.4.3, eigen coupling factors play a significant role in the design of maximum bandwidth filters. On the other hand, we may be interested in minimizing rather than maximizing bandwidth. In this case, we must make the resonator loss as low as possible, so as to maximize the Q; consequently, the loss eigenstate having the lowest eigen loss tangent now represents the ideal. In other words, it now becomes desirable to determine the electrode configuration and sample shape which excites a mode resembling, point by point, the loss eigenstate associated with this lowest eigen loss tangent.

Unlike the case of coupling factors, loss is, on physical grounds, clearly invariant to change of independent variables. Consequently, the loss eigenstates (including the optimum one) are not beset with the indeterminacy appearing in the evaluation of the coupling eigenstates.

As stated in Section 1.4.5, we should note that the smallest eigen loss tangent may be considerably smaller than the smallest loss tangent of any element of $[M]$. An interpretation may be drawn from the comment near the end of Section 1.3 that the loss tangent of off-diagonal elements of $[M]$ may correspond to power *gain*, not loss, if the components of $[F]$ obey certain phase relations. In other words, it is possible for the effects of one loss tangent to cancel in part those of another.

For example, Figure 1.3 indicates that it should be possible to build PZT 65/35 resonators having a Q of 1000 in the fully poled state with the excitation constrained to obey Equation 1.140. This may appear surprising as the elastic $Q = 1/\delta_+$ indicated by Figure 1.2 is only 160. In practice, though, by judicious choice of electroding, it actually turns out to be quite easy to obtain a Q of at least 240 without violating Equation 1.140 in PZT 65/35 resonators.

1.4.7. Eigen Coupling Factors; the Lossy, AC Case

In the lossy, ac case the coefficient matrices $[M^0]$ and $[M^1]$ become complex, as a comparison of Equations 1.61 and 1.121 indicates. However, U_D, U_E, and U_M are only associated with the real part of $[M]$. Consequently, in extending the eigencoupling concept to the lossy, ac case, one only uses the real part of $[M]$ in forming $[M^0]$ and $[M^1]$.

Other minor modifications necessary to transform the dc theory of invariant coupling to the ac theory include adding appropriate factors of $\frac{1}{2}$ in Equations 1.53, 1.56, 1.59, 1.64, 1.80, 1.81, etc. Transposed vectors are changed to Hermitian conjugate vectors in Equations 1.64, 1.69 to 1.81, etc. As the $[K^{(A)}]$ will remain pure real, there is no difference between the transpose and the Hermitian conjugate of $[\mathscr{K}]$.

2 The Analysis of Multielectrode Piezoelectric Plates; Green's Function Techniques

Certain types of complex piezoelectric devices can be very conveniently designed or analyzed by the use of Green's functions. All varieties of multielectrode, thin-wafer, contour extensional resonators fall into this category. (See Figure 2.1 for some typical examples.) One can also treat multilayer plates by means of Green's functions.

In the following chapters, we will describe procedures for dealing with device configurations that fall outside the scope of Green's function techniques. However, the number of these configurations that defy Green's function analysis and yet are practically significant is surprisingly small. This is especially true if one excludes the "energy-trapping" devices discussed in Chapter 5.

The basic Green's function philosophy we will present here is fairly simple. Consider the thin-wafer and electrode configurations shown in Figure 2.1. (The unshown face of each wafer will be assumed to have an electrode pattern identical to that of the visible face.) It is well known that the mechanical displacement in these wafers obeys a homogeneous wave equation except in the vicinity of the electrode edges.[40,41,42,43] Early theories of these devices consisted of finding solutions to the homogeneous wave equation under each electrode and then of splicing the resulting solutions at the electrode edges by matching boundary conditions. Examples of this correct but cumbersome technique may be found in articles by Berlincourt *et al.*[41] (the thin bar with a single electrode on each face), van der Veen[44] (the thin bar with two equal electrodes on each face), Mason[40] (the thin disk

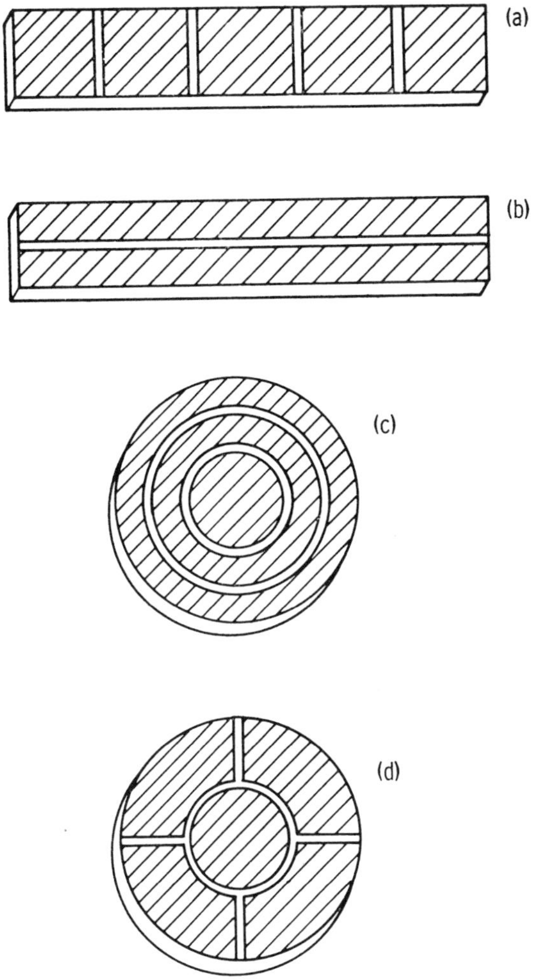

Figure 2.1. Typical electrode configurations for thin piezoelectric wafers. The bottom faces of the wafers are understood to have the same electrode configuration as the faces shown.

with a single electrode on each face), and Munk[45] (the thin disk with two concentric electrodes on each face).

A more satisfactory way of dealing with this problem, however, is to realize that the inhomogeneous term arising at the edge of each electrode in the wave equation is a force density function. Physically it is this force density that causes the material to move. As these forces act only at the electrode edges, we could, in effect, say that electromechanical energy conversion occurs only there, at the electrode edges. Using a conventional Green's function technique,[46,47] we may solve the inhomogeneous wave equation by integrating the inhomogeneous term over the surface of the entire sample.

This procedure was first introduced by Jacobsen[48] in a brief article on quartz microwave transducers. Apparently at least five years then passed before additional use was made of Green's functions in piezoelectric device design. This seems rather sad, as those were years in which piezoelectric devices rapidly became more complicated, and the Green's function technique in many respects renders the mathematical design complexity independent of the actual physical complexity of the hardware.

2.1. Equivalent Circuit and Admittance Matrix of a Slim Multielectrode Bar

2.1.1. Theory

Let us now demonstrate the mechanics of piezoelectric Green's function analysis by developing an equivalent circuit for the N-electrode-pair ferroelectric ceramic bar shown in Figure 2.2. The manner in which currents, voltages, forces, and coordinates are defined is illustrated in that figure. It will be assumed that cross-sectional dimensions (w,τ) are small, so that T_1 is the only stress and electrical fringing is negligible. Thus, x_2 and x_3 are unused coordinates. We will analyze the problem under sinusoidal steady-state conditions; relevant variables may be regarded as containing a suppressed $e^{j\omega t}$ dependence. Under these assumptions, the constituent relations, Equations 1.18,

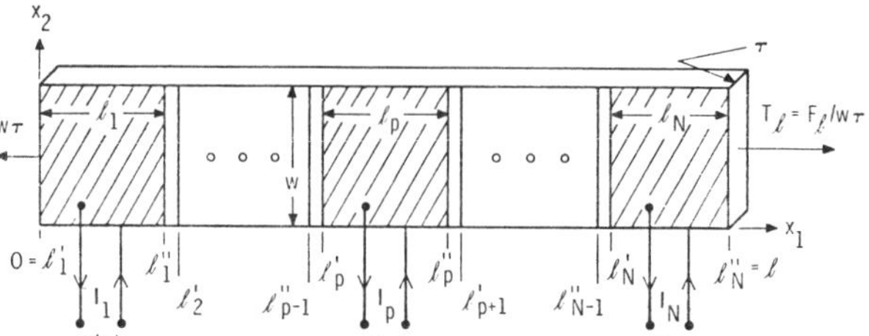

Figure 2.2. The N-electrode-pair bar problem. The bottom face is understood to have the same electrode configuration as the face shown.

reduce to

$$\begin{aligned} S_1 &= s_{11}^E T_1 + d_{31} E_3 \\ D_3 &= d_{31} T_1 + \epsilon_{33}^T E_3 \end{aligned} \tag{2.1}$$

In the electroded regions, E_3 is ϕ/τ, where ϕ is the voltage applied at the electrode pair in question (see Figure 2.2). The coefficients s_{11}^E, d_{31}, and ϵ_{33}^T are assumed to be complex, as discussed in Section 1.3, in order to represent the most general phenomenological loss. As the bar has been assumed to be ferroelectric, different electrode regions may be in different polarization states. Thus s_{11}^E, d_{31}, and ϵ_{33}^T may vary in the different electroded regions, although only d_{31} may show greater than about 20 percent variation from electrode to electrode.[23,49]

In the unelectroded regions, D_3 will necessarily be zero, as no free charge can come or go in the absence of an electrode. Consequently, in these regions, Equations 2.1 reduce to

$$S_1 = \left(s_{11}^E - \frac{d_{31}^2}{\epsilon_{33}^T} \right) T_1 = s_{11}^D T_1 \tag{2.2}$$

Here s_{11}^D is the compliance evaluated at constant D_3 — see Equations 1.21. The coefficient s_{11}^D may differ from s_{11}^E by as much as 20 percent.[49]

Equations 2.1 and 2.2 may be combined into a single set of equations approximately valid over the entire bar except possibly at the ends:

$$\begin{aligned} S_1 &= s_{11} T_1 + d_{31} E_3' \\ D_3 &= d_{31} T_1 + \epsilon_{33}^T E_3 \end{aligned} \tag{2.3}$$

The following conventions apply here:

d_{31} and ϵ_{33}^T are functions of x_1

$$E_3' = \begin{cases} \phi_p/\tau \text{ at electrode pair } p \\ 0 \text{ in unelectroded regions} \end{cases}$$

s_{11} is a constant over x_1, and is some appropriate average of s_{11}^D and s_{11}^E defined so as to make Equation 2.3 yield the same resonant frequencies as the exact equations.

Several fairly obvious ways exist to specify s_{11}; one of these is described in detail in the Appendix to Reference 42.

It should be noted that in the approximation of no fringing, E_3 and E_3' are identical in the electroded regions. However, they differ by $d_{31}T_1/\epsilon_{33}^T$ in the unelectroded regions.

Substitution of an average compliance s_{11} for the exact, x_1-dependent value constitutes an approximation that does not affect our accuracy significantly but greatly simplifies subsequent derivations. The relatively weak dependence of the exact compliance on x_1 makes feasible its approximation by a constant.

Newton's equation of momentum conservation in differential form is

$$\frac{\partial T_1}{\partial x_1} = -\rho\omega^2 u_1 \tag{2.4}$$

where ρ is the material density. If we combine this and the definition of strain Equation 1.2 with Equation 2.3, we find

$$\frac{\partial^2 u_1}{\partial x_1^2} + s_{11}\rho\omega^2 u_1 = \frac{\partial(d_{31}E_3')}{\partial x_1} \tag{2.5}$$

The right side of Equation 2.5 represents the inhomogeneous component of the wave equation discussed at the beginning of this chapter. In view of the electrode configuration illustrated in Figure 2.2, the definition of E_3' makes this component a set of singularity functions, one acting at each electrode edge:

$$\frac{\partial(d_{31}E_3')}{\partial x_1} = \sum_{p=1}^{N} \frac{d_{31}(p)\phi_p}{\tau}[\delta(x_1 - l_p') - \delta(x_1 - l_p'')] \tag{2.6}$$

The symbol $d_{31}(p)$ represents d_{31} evaluated at electrode pair p.

Early workers treated Equation 2.5 by putting these singularities in as boundary conditions. However, in the Green's function formulation, they are merely integrated simultaneously (with appropriate weights) over the entire bar.

Having spent several pages discussing the Green's function in vague terms, it is now time for us to get down to specifics. Let a force of value $(w\tau)/s_{11}$ at angular frequency ω be applied to the bar in the x_1 direction at some point x''. Let the bar have no voltages and no other forces applied, and let the ends be free. We then define the Green's function evaluated at some other point x' to be the resulting particle displacement at that point. Mathematically, this means the Green's function $G(x' \mid x'')$ obeys the differential equation

$$\left(\frac{\partial^2}{\partial x'^2} + s_{11}\rho\omega^2\right)G(x' \mid x'') = -\delta(x' - x'') \tag{2.7}$$

and the boundary conditions

$$0 = s_{11}T_1 + d_{31}E_3'\big|_{x'=0,l} = S_1\big|_{x'=0,l} = \frac{\partial G(x' \mid x'')}{\partial x'}\bigg|_{x'=0,l} \tag{2.8}$$

Since $G(x' \mid x'')$ obeys a second-order differential equation, it is completely specified by these two boundary conditions.

A Fourier expansion solution for $G(x' \mid x'')$ may be found quite easily. Assume

$$G(x' \mid x'') = A_0(x'') + \sum_{\nu=1}^{\infty} A_\nu(x'') \cos(\nu\pi x'/l) \tag{2.9}$$

A termwise examination indicates that this series obeys the boundary conditions of Equation 2.8.

The $A_\nu(x'')$ coefficients are found by substituting the series into the differential equation, Equation 2.7, multiplying the resulting equation on both sides by $\cos(\mu\pi x'/l)$, and integrating from 0 to l:

$$G(x' \mid x'') = -\frac{1}{\omega^2 \rho s_{11} l} - \sum_{\nu=1}^{\infty} \frac{2}{l} \frac{\cos(\nu\pi x'/l)\cos(\nu\pi x''/l)}{\omega^2 \rho s_{11} - \nu^2\pi^2/l^2} \tag{2.10}$$

We note that $G(x' \mid x'')$ is symmetric in x' and x''. This symmetry of the Green's function is characteristic of elasticity problems and is, in fact, a consequence of the reciprocity theorem as applied to mechanical systems.[50] In spite of the previous definition of x' and x'', we shall call x' the source coordinate and x'' the observer coordinate. This convention makes subsequent results more intuitive. Moreover, since $G(x' \mid x'')$ is symmetric, this convention is not actually inconsistent with the original definitions.

The wave equation may now be integrated for u_1. Multiply Equation 2.7 by $u_1(x')$ and Equation 2.5 by $G(x' \mid x'')$. Subtract the second equation from the first, and integrate the result from ϵ to $l - \epsilon$, where ϵ is an infinitesimal distance:

$$\int_\epsilon^{l-\epsilon} \left[u_1(x')\left(\frac{\partial^2 G(x' \mid x'')}{\partial x'^2} + s_{11}\rho\omega^2 G(x' \mid x'')\right) - G(x' \mid x'')\left(\frac{\partial^2 u_1(x')}{\partial x'^2} + s_{11}\rho\omega^2 u_1(x')\right)\right] dx'$$

$$= \int_\epsilon^{l-\epsilon} \left[u_1(x')\{-\delta(x' - x'')\} - G(x' \mid x'')\frac{\partial(d_{31}E_3')}{\partial x'}\right] dx' \tag{2.11}$$

Two terms in the first integral cancel; the other two may be integrated by parts. Performing these operations, we have an explicit expression for u_1 at the observer coordinate:

$$u_1(x'') = -\int_{\epsilon}^{l-\epsilon} G(x' \mid x'') \frac{\partial(d_{31}E_3')}{\partial x'}\, dx' - \left[\frac{\partial G(x' \mid x'')}{\partial x'} u_1(x') - \frac{\partial u_1(x')}{\partial x'} G(x' \mid x'')\right]_{\epsilon}^{l-\epsilon} \tag{2.12}$$

In view of the boundary conditions of Equation 2.8 on $G(x' \mid x'')$, the first term of the above limit is zero. The second may be expressed as

$$\begin{aligned}\frac{\partial u_1(x')}{\partial x'} G(x' \mid x'')\Big|_{\epsilon}^{l-\epsilon} &= S_1(x')G(x' \mid x'')\Big|_{\epsilon}^{l-\epsilon} \\ &= \frac{s_{11}}{w\tau}[F_l G(l \mid x'') - F_0 G(0 \mid x'')] + d_{31}E_3' G(x' \mid x'')\Big|_{\epsilon}^{l-\epsilon} \\ &= \frac{s_{11}}{w\tau}[F_l G(l \mid x'') - F_0 G(0 \mid x'')] \\ &\quad - \left(\int_{-\epsilon}^{\epsilon} + \int_{l-\epsilon}^{l+\epsilon}\right) G(x' \mid x'') \frac{\partial(d_{31}E_3')}{\partial x'}\, dx' \end{aligned} \tag{2.13}$$

The final step in Equation 2.13 is valid because $d_{31}E_3'$ varies very rapidly at $x' = 0, l$, while $G(x' \mid x'')$ does not. Thus, our formula for the displacement at the point of observation becomes a source coordinate system integral of the inhomogeneous or driving terms:

$$u_1(x'') = -\int_{-\epsilon}^{l+\epsilon} G(x' \mid x'') \frac{\partial(d_{31}E_3')}{\partial x'}\, dx' + \frac{s_{11}}{w\tau}[F_l G(l \mid x'') - F_0 G(0 \mid x'')] \tag{2.14}$$

If Equation 2.6 for the derivative of $(d_{31}E_3')$ and Equation 2.10 for the Green's function are substituted into Equation 2.14, we get the response for $u_1(x'')$ in terms of the input forces and voltages and the properties of the bar:

$$\begin{aligned} u_1(x'') &= \frac{2}{\rho s_{11}\tau l}\sum_{p=1}^{N}\sum_{\nu=1}^{\infty}\frac{\phi_p d_{31}(p)[\cos(\nu\pi l_p'/l) - \cos(\nu\pi l_p''/l)]\cos(\nu\pi x''/l)}{\omega^2 - \nu^2\pi^2/(\rho s_{11}l^2)} \\ &\quad + \frac{2}{\rho l w\tau}\sum_{\nu=1}^{\infty}\frac{[F_0 - (-1)^\nu F_l]\cos(\nu\pi x''/l)}{\omega^2 - \nu^2\pi^2/(\rho s_{11}l^2)} + \frac{1}{\rho l w\tau}\frac{(F_0 - F_l)}{\omega^2} \end{aligned} \tag{2.15}$$

This formula has an interesting physical interpretation. The last term represents the uniform or center-of-mass motion of the bar due to the net force applied at the ends. The cos $(\nu\pi x''/l)$ factor in each term of the two series is the displacement function associated with the νth length extensional mode of the sample. The denominator of those terms is the resonance factor associated with the νth mode. The remaining factors in each term indicate how efficiently the various forces and voltages are coupled to the νth mode.

It is interesting to devote a bit of special attention to the denominator or resonance factors of Equation 2.15. The denominator of term ν is zero at the complex resonant frequency of mode ν. If we represent this νth complex resonant frequency along with its real and imaginary parts by

$$\omega_\nu = {}^R\omega_\nu + j\,{}^I\omega_\nu = \nu\pi/(l\sqrt{\rho s_{11}}) \tag{2.16}$$

and similarly represent the complex average compliance (see Equation 1.33) by

$$s_{11} = {}^R s_{11} - j\,{}^I s_{11} \tag{2.17}$$

we find the Q of the νth mode is

$$Q_\nu = \frac{{}^R\omega_\nu}{2\,{}^I\omega_\nu} = \frac{{}^R s_{11}}{{}^I s_{11}} \tag{2.18}$$

In Equation 2.18, terms on the order of $1/Q_\nu^2$ are neglected. If the loss tangent of s_{11} does not depend on frequency, the Q of all the mechanical modes of the system are thus equal.*

The closest a real driving frequency can approach the νth complex resonance is $\omega = {}^R\omega_\nu$. Since the propagation velocity of sound in the bar is $v = ({}^R s_{11}\rho)^{-1/2}$, this will occur when the bar is exactly ν acoustic half wavelengths long at the driving frequency:

$${}^R\omega_\nu = \frac{\nu\pi v}{l} = \frac{2\pi v}{\lambda_\nu} \tag{2.19}$$

Let us now complete the evaluation of the admittance matrix for the bar shown in Figure 2.2. The current at electrode pair p is the time derivative of the electric displacement integrated over that electrode

* It can be shown, however, on purely mathematical grounds that the loss tangent must be frequency dependent unless it is zero.

pair. In view of Equation 2.3 and the definition of strain, this becomes

$$I_p = \frac{j\omega w l_p}{\tau} \epsilon_{33}^{S_1}(p)\phi_p + \frac{j\omega w d_{31}(p)}{s_{11}} [u_1(l_p'') - u_1(l_p')] \tag{2.20}$$

where $\epsilon_{33}^{S_1}(p)$ is the dielectric constant in electrode region p evaluated with motion prevented in the x_1-direction,

$$\epsilon_{33}^{S_1}(p) = \epsilon_{33}^{T}(p) - d_{31}(p)^2/s_{11} \tag{2.21}$$

If Equation 2.15 for u_1 is substituted into Equation 2.20, we get

$$I_p = j\omega C_0(p)\phi_p - \frac{2j\omega w d_{31}(p)}{\rho l \tau s_{11}^2}\left\{\frac{s_{11}}{w}\sum_{\nu=1}^{\infty}\frac{A_{\nu p}[F_0 - (-1)^\nu F_l]}{\omega^2 - \nu^2\pi^2/(\rho s_{11} l^2)} + \sum_{q=1}^{N}\sum_{\nu=1}^{\infty}\frac{\phi_q d_{31}(q) A_{\nu p} A_{\nu q}}{\omega^2 - \nu^2\pi^2/(\rho s_{11} l^2)}\right\} \tag{2.22}$$

where $C_0(p)$ is the so-called "clamped capacitance" of electrode pair p,

$$C_0(p) = l_p w \epsilon_{33}^{S_1}(p)/\tau \tag{2.23}$$

The quantity $A_{\nu p}$ is a measure of the coupling between electrode pair p and mode ν,

$$A_{\nu p} = \cos(\nu\pi l_p'/l) - \cos(\nu\pi l_p''/l) \tag{2.24}$$

Now differentiation of I_p with respect to ϕ_q in Equation 2.22 gives the electrical portion of the desired admittance matrix,

$$\frac{\partial I_p}{\partial \phi_q} = Y_{pq} = j\omega C_0(p)\delta_{pq} - \frac{j\omega w d_{31}(p) d_{31}(q)}{l \rho s_{11}^2 \tau}\sum_{\nu=1}^{\infty}\frac{2A_{\nu p}A_{\nu q}}{\omega^2 - \nu^2\pi^2/(\rho s_{11} l^2)} \tag{2.25}$$

All quantities on the right side of this equation may be evaluated from a knowledge of the sample material coefficients. The physical interpretation of the last term is that the admittance between electrode pair p and electrode pair q is a sort of "communication" with the mechanical eigenmodes of the system acting as a "medium" for that communication. How good that communication is by means of mode ν depends on both $A_{\nu p}$ and $A_{\nu q}$, which are measures of how well electrode pairs p and q, respectively, are coupled to mode ν.

If we differentiate I_p with respect to F_0 or F_l, we obtain the electromechanical portion of the admittance matrix:

$$\begin{aligned}\frac{\partial I_p}{\partial F_0} &= Y_{p0} = -\frac{2j\omega d_{31}(p)}{\rho\tau l s_{11}}\sum_{\nu=1}^{\infty}\frac{A_{\nu p}}{\omega^2 - \nu^2\pi^2/(\rho s_{11}l^2)}\\ \frac{\partial I_p}{\partial F_l} &= Y_{pl} = \frac{2j\omega d_{31}(p)}{\rho\tau l s_{11}}\sum_{\nu=1}^{\infty}\frac{(-1)^\nu A_{\nu p}}{\omega^2 - \nu^2\pi^2/(\rho s_{11}l^2)}\end{aligned} \tag{2.26}$$

The same result could be obtained by differentiating $-\dot{u}_1(0)$ and $\dot{u}_1(l)$, the particle velocities at the ends of the bar as evaluated from Equation 2.15, with respect to ϕ_p:

$$\begin{aligned}-\frac{\partial \dot{u}_1(0)}{\partial \phi_p} &= Y_{0p} = Y_{p0}\\ \frac{\partial \dot{u}_1(l)}{\partial \phi_p} &= Y_{lp} = Y_{pl}\end{aligned} \tag{2.27}$$

These electromechanical admittances are especially applicable to the field of transducer design.

Finally, if we differentiate the expressions for $-\dot{u}_1(0)$ and $\dot{u}_1(l)$ as obtained from Equation 2.15 with respect to F_0 and F_l, we arrive at the mechanical portion of the admittance matrix:

$$\begin{aligned}-\frac{\partial \dot{u}_1(0)}{\partial F_0} &= Y_{00} = \frac{\partial \dot{u}_1(l)}{\partial F_l} = Y_{ll}\\ &= -\frac{j\omega}{\rho l w\tau}\left(2\sum_{\nu=1}^{\infty}\frac{1}{\omega^2 - \nu^2\pi^2/(\rho s_{11}l^2)} + \frac{1}{\omega^2}\right)\\ -\frac{\partial \dot{u}_1(0)}{\partial F_l} &= Y_{0l} = \frac{\partial \dot{u}_1(l)}{\partial F_0} = Y_{l0}\\ &= \frac{j\omega}{\rho l w\tau}\left(2\sum_{\nu=1}^{\infty}\frac{(-1)^\nu}{\omega^2 - \nu^2\pi^2/(\rho s_{11}l^2)} + \frac{1}{\omega^2}\right)\end{aligned} \tag{2.28}$$

This completes the derivation of the admittance matrix for the sample shown in Figure 2.2. It is, of course, impossible to represent this type of distributed medium problem over all frequencies by a finite number of lumped R's, L's, and C's. However, in the vicinity of some particular mode, the admittance matrix given by Equations 2.25 to 2.28 may be described pictorially by an extremely simple model. In particular, let the frequency be such that only the νth term in Equations 2.25

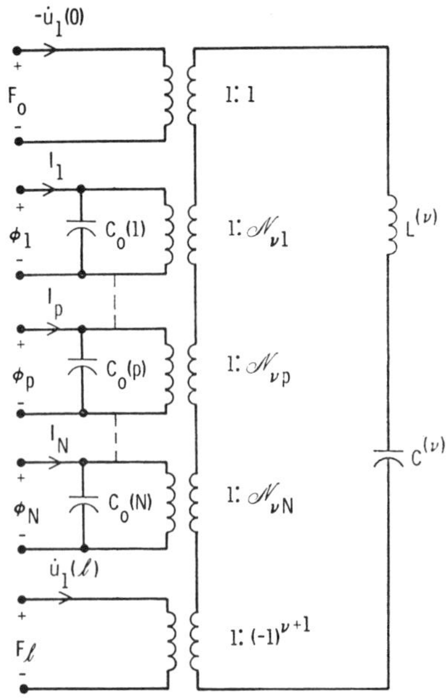

Figure 2.3. Equivalent circuit near resonance ν for the N-electrode-pair ferroelectric ceramic bar shown in Figure 2.2.

to 2.28 is significant. If we then define the following lumped circuit elements,

$$
\begin{aligned}
L^{(\nu)} &= \frac{\rho w l \tau}{2} = \tfrac{1}{2} \text{ mass of bar} \\
&= \text{equivalent motional inductance} \\
C^{(\nu)} &= \frac{2}{\pi^2} \frac{l s_{11}}{w \tau \nu^2} = \text{equivalent motional capacitance} \\
\mathcal{N}_{\nu p} &= \frac{w d_{31}(p)}{s_{11}} A_{\nu p} = \text{equivalent electromechanical transformer turns ratio at electrode } p
\end{aligned}
\tag{2.29}
$$

it is possible to represent our piezoelectric bar by the equivalent circuit shown in Figure 2.3. Note that $C^{(\nu)}$ and $\mathcal{N}_{\nu p}$ are complex, but $L^{(\nu)}$ is pure real.

To summarize, the fundamental concept introduced in this section is that electromechanical energy conversion takes place only at the electrode edges rather than under the entire electrode. This is a highly nonintuitive and significant result. A skeptical reader may well need experimental evidence to be convinced of the truth of this idea. In this context, it is possible to describe two classes of observations.

First, while no direct quantitative evidence exists to verify the concept in question, Equation 2.25 depends on it very basically, and this equation has been checked many times against experiment with excellent agreement (see Reference 42).

Second, when large electric fields are applied to a ferroelectric ceramic plate to change the polarization state in some electrode region, cracks occasionally develop. These cracks are usually localized at the electrode edges — where the present theory indicates the electrically induced mechanical forces should act.

2.1.2. Device Applications

Multielectrode piezoelectric bars have a number of device applications, either actually developed or potentially possible. The most straightforward of these, based on the equivalent circuit of Figure 2.3, is as miniaturized tuned transformers. For example, the acoustic wavelength of most piezoelectric materials is such that one should be able to replace the 455 kHz tuned IF transformers in an AM receiver by a two-electrode-pair piezoelectric bar somewhat less than 1 cm in length.[133]

Three-electrode-pair ferroelectric ceramic bars have been proposed by Land[6,51] for use in electrically tuned oscillators and FM discriminators.

An electrically tuned oscillator of this type is shown in Figure 2.4. The frequency of oscillation in this case will be at the resonance of the

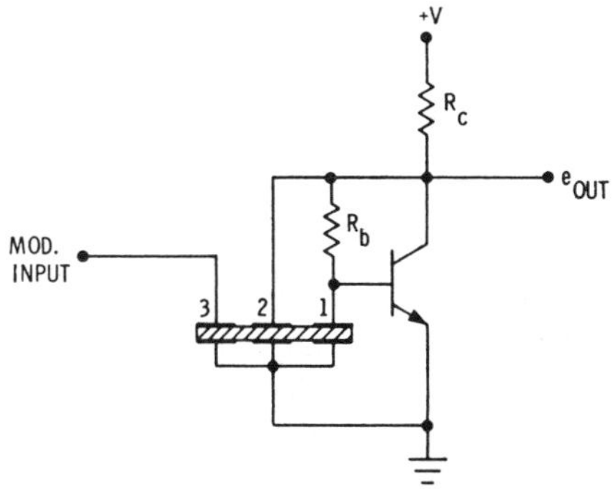

Figure 2.4. An electrically tuned oscillator with a three-electrode-pair ferroelectric ceramic bar for frequency control in the feedback circuit.

three-electrode-pair bar in the feedback circuit. As we shall now demonstrate, this resonance itself may be electrically shifted.

Assume electrode region 3 of the bar is connected to a high impedance charge source (labeled "Mod. input" in Figure 2.4), which can change the polarization state of the ceramic at that electrode pair. Let us now reflect the clamped capacitance $C_0(3)$ of that electrode region across its electromechanical transformer $\mathcal{N}_{\nu 3}$, so that the general equivalent circuit of Figure 2.3 becomes specialized to the present case as shown in Figure 2.5. In Figure 2.5, note that this reflected

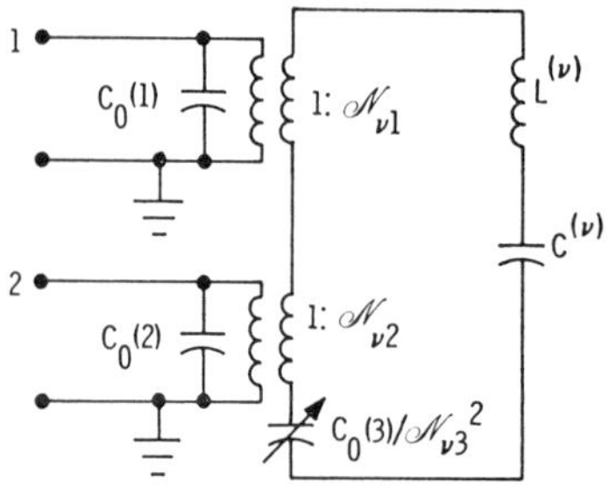

Figure 2.5. Equivalent circuit for the three-electrode-pair ferroelectric ceramic bar with polarization of electrode Region 3 variable.

capacitance $C_0(3)/\mathcal{N}_{\nu 3}^2$ now appears in series with the frequency-determining series resonant circuit and raises the overall resonant frequency. Moreover, varying the polarization at electrode pair 3 will alter the piezoelectric constant there, $d_{31}(3)$, and hence, according to Equation 2.29, will alter the value of this reflected capacitance. For many ferroelectric ceramics, the possible range of $\mathcal{N}_{\nu 3}$ in relation to $C^{(\nu)}$ and $C_0(3)$ is such that five percent or more resonant frequency shifts can be obtained in this way. Analogous variable-frequency disk resonators (see Section 2.2) can increase this figure to 15 percent, because piezoelectric effects in disks are inherently stronger than in bars.

The earlier-mentioned FM discriminator circuit is shown in Figure 2.6 and requires two identical three-electrode-pair bars. Electrode region 3 of the top bar is short-circuited, while electrode region 3 of the bottom bar is open-circuited. For the same reasons described above in discussing the variable frequency oscillator, this introduces an additional series capacitance in the equivalent circuit for the bottom

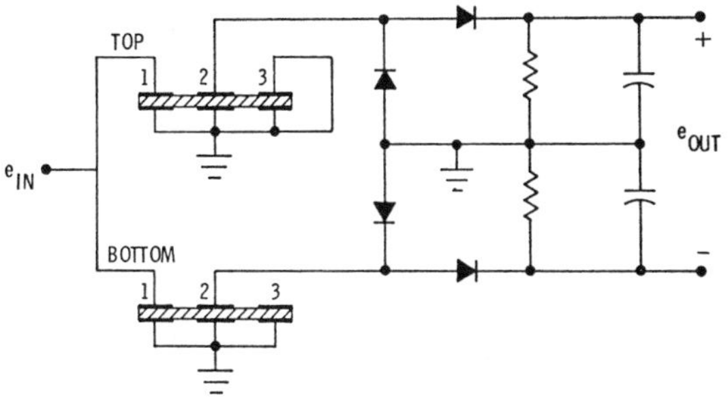

Figure 2.6. FM discriminator circuit employing 2 three-electrode-pair piezoelectric bars.

bar. Consequently, the bottom bar will have a higher resonant frequency than the top bar. Thus, the Y_{12} transfer admittance of the top bar will peak at a lower frequency than the Y_{12} transfer admittance of the bottom bar, as illustrated in Figure 2.7. The *RC*-diode network to the right of the bars acts as a detector whose output is proportional to the difference signal at electrode pair 2 of the two bars. Hence the output voltage at the capacitors of this circuit will bear the *S*-curve relation to input frequency shown in Figure 2.7, which is required to produce FM discrimination.

Additional device applications of multielectrode ferroelectric ceramic bars lie in nondestructive readout memory devices[4,5] and logic elements.[5] However, these devices may be considerably more efficiently designed in asymmetric disk than bar geometry, and we shall thus defer their discussion to Section 2.3.

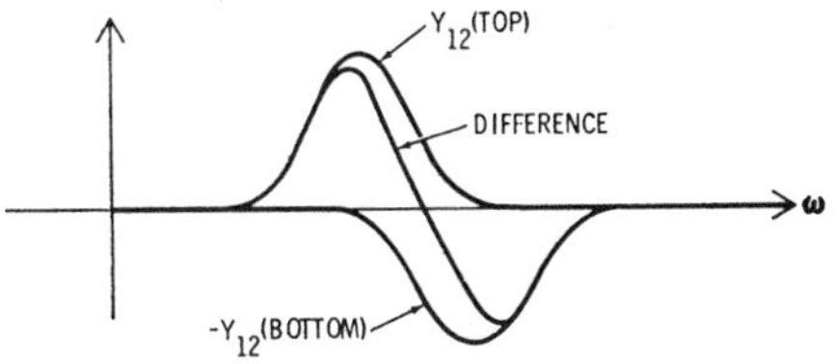

Figure 2.7. Y_{12} transfer admittances for top and bottom three-electrode-pair piezoelectric bars in the FM discriminator circuit, and difference curve between the two. The overall transfer function for the discriminator circuit is proportional to this *S*-shaped difference curve.

2.2. Equivalent Circuit and Admittance Matrix of a Symmetric Multielectrode Disk

2.2.1. Theory

We shall now take up the thin piezoelectric ceramic disk with N concentric electrode pairs shown in Figure 2.8. Currents, forces, and voltages are defined there by illustration. Let us assume that the thickness τ is small so that electrical fringing is negligible and that there are no stress components normal to the plane of the disk. As in the previous section, time dependence will be restricted to $e^{j\omega t}$.

Under these assumptions, a Green's function and eigenmode analysis of the admittance matrix and equivalent circuit may be carried out precisely as in the previous section on bars. However, details such as the appearance of covariant derivatives and Bessel functions make the actual mechanics of disk analysis quite tedious. Consequently, we shall here just quote the results of this analysis which has been published step by step elsewhere.[43] Some of the earlier-mentioned complications,

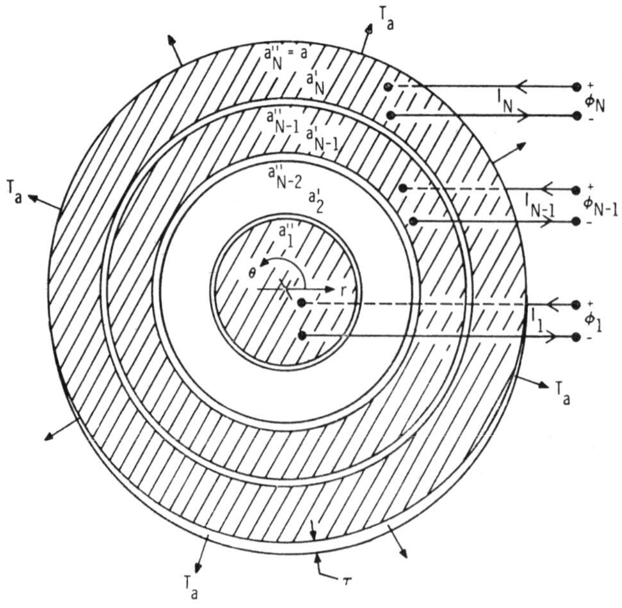

Figure 2.8. The N-electrode-pair, symmetric, ferroelectric ceramic disk. It is assumed the bottom face is electroded identically to the visible face.

such as the use of covariant derivatives, are treated extensively in the following section.

In particular, the electrical portion of the admittance matrix for this disk is given by

$$\frac{\partial I_p}{\partial \phi_q} = Y_{pq} = j\omega C_0(p)\delta_{pq} - \frac{4\pi j\omega d_{31}(p)d_{31}(q)}{\rho\tau(s_{11} + s_{12})^2} \times \sum_{\nu=1}^{\infty} \frac{K_\nu^2 B_{\nu p} B_{\nu q}}{(\omega^2 - \omega_\nu^2)(K_\nu^2 - 1 + \sigma^2)} \tag{2.30}$$

This formula is analogous to Equation 2.25 for bars. In this equation the compliances s_{11} and s_{12} are constants over the disk and represent averaged values of s_{11}^D and s_{11}^E, and s_{12}^D and s_{12}^E, respectively.[43] In addition, σ is an averaged Poisson's ratio,

$$\sigma = -s_{12}/s_{11} \tag{2.31}$$

The quantity $C_0(p)$ is the capacitance of electrode pair p with motion prevented in the plane of the disk,

$$C_0(p) = \frac{\pi(a_p''^2 - a_p'^2)}{\tau} \epsilon_{33}^{S_2}(p) \tag{2.32}$$

where $\epsilon_{33}^{S_2}(p)$ is the dielectric constant evaluated with motion prevented in the plane of the disk,

$$\epsilon_{33}^{S_2}(p) = \epsilon_{33}^{T}(p) - \frac{2d_{31}(p)^2}{(s_{11} + s_{12})} \tag{2.33}$$

These definitions for $C_0(p)$ and $\epsilon_{33}^{S_2}(p)$ should be compared with the corresponding definitions for bars, Equations 2.23 and 2.21. The symbol $B_{\nu p}$ is analogous to $A_{\nu p}$ of Equation 2.24 and is a measure of the coupling between electrode pair p and the νth angularly symmetric acoustic mode of the disk,

$$B_{\nu p} = \frac{a_p'' J_1(a_p'' K_\nu/a) - a_p' J_1(a_p' K_\nu/a)}{a J_1(K_\nu)} \tag{2.34}$$

Graphs are available which are useful in the numerical computation of this function (see Reference 43).

The resonant frequencies of the disk ω_ν appearing in Equation 2.30 and the so-called eigenvalues of the problem K_ν are related by

$$K_\nu = a\omega_\nu[\rho s_{11}(1 - \sigma^2)]^{1/2} \tag{2.35}$$

where K_ν is the νth solution of

$$(1 - \sigma)J_1(K_\nu) = K_\nu J_0(K_\nu) \tag{2.36}$$

Solutions of this equation are given in Table 2.1 for various values of ν and σ. Also given in Table 2.1 is $(\nu - \frac{1}{4})\pi$, which K_ν asymptotically approaches irrespective of σ as ν increases.

We shall consider the circumference of the disk to be the mechanical terminal in this case, just as we considered the ends of the bar to be mechanical terminals in the previous section. The variable of this terminal which is analogous to current is radial particle velocity

Table 2.1. Solutions of $(1 - \sigma)J_1(K_\nu) = K_\nu J_0(K_\nu)$

ν	$\sigma = 0.20$	$\sigma = 0.26$	$\sigma = 0.32$	$\sigma = 0.38$	$\sigma = 0.44$	$(\nu - \frac{1}{4})\pi$
1	1.9844	2.0235	2.0612	2.0974	2.1322	2.3562
2	5.3701	5.3816	5.3931	5.4046	5.4160	5.4978
3	8.5600	8.5671	8.5742	8.5813	8.5883	8.6394
4	11.7232	11.7283	11.7335	11.7386	11.7437	11.7810
5	14.8770	14.8811	14.8851	14.8892	14.8932	14.9226
6	18.0266	18.0299	18.0333	18.0366	18.0400	18.0641
7	21.1738	21.1766	21.1795	21.1823	21.1851	21.2057

$\dot{u}_r(a)$. Let a uniform radial stress $T_{rr} = T_a$ act on the edge of this disk. Then the mechanical variable which corresponds to voltage is most conveniently chosen as

$$F_a = 2\pi a \tau T_a \tag{2.37}$$

In a sense, F_a represents the total force on the disk, in that it is stress times the area the stress acts on.

The electromechanical portion of the disk admittance matrix is given as the derivative of I_p with respect to F_a or alternatively as the derivative of $\dot{u}_r(a)$ with respect to ϕ_p:

$$\begin{aligned} \frac{\partial I_p}{\partial F_a} &= \frac{\partial \dot{u}_r(a)}{\partial \phi_p} = Y_{ap} \\ &= -\frac{j\omega d_{31}(p)}{\rho\tau a(s_{11} + s_{12})} \sum_{\nu=1}^{\infty} \frac{2K_\nu^2 B_{\nu p}}{(\omega^2 - \omega_\nu^2)(K_\nu^2 - 1 + \sigma^2)} \end{aligned} \tag{2.38}$$

Finally, the mechanical portion of the admittance matrix is

$$\frac{\partial \dot{u}_r(a)}{\partial F_a} = Y_{aa} = \frac{-j\omega}{\pi a^2 \rho\tau} \sum_{\nu=1}^{\infty} \frac{K_\nu^2}{(\omega^2 - \omega_\nu^2)(K_\nu^2 - 1 + \sigma^2)} \tag{2.39}$$

The admittance matrix of the disk, Equations 2.30, 2.38, and 2.39, may be represented by an equivalent circuit which is very similar to that used to characterize the N-electrode-pair bar. In particular, if frequency is chosen so that only the νth term of the various infinite series contributes significantly, the disk may be described by the circuit shown in Figure 2.9. The lumped component elements appearing in

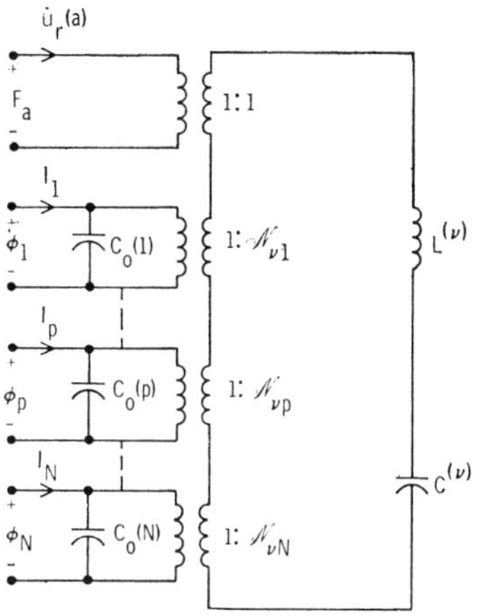

Figure 2.9. Equivalent circuit near resonance ν for the symmetric N-electrode-pair ferroelectric ceramic disk shown in Figure 2.8.

that circuit are related to the properties of the disk by the definitions

$$L^{(\nu)} = M\left(1 - \frac{1-\sigma^2}{K_\nu^2}\right) = \text{equivalent motional inductance of mode } \nu \,(M = \text{mass of disk} = \pi a^2 \tau \rho)$$

$$C^{(\nu)} = \frac{s_{11}(1-\sigma^2)}{\pi\tau(K_\nu^2 - 1 + \sigma^2)} = \text{equivalent motional capacitance of mode } \nu$$

$$\mathcal{N}_{\nu p} = \frac{2\pi a d_{31}(p) B_{\nu p}}{(s_{11} + s_{12})} = \text{equivalent electromechanical transformer turns ratio at electrode } p. \tag{2.40}$$

If loss is represented by a complex s_{11} but real σ, then $C^{(\nu)}$ and $\mathcal{N}_{\nu p}$ will be complex while $L^{(\nu)}$ remains pure real. However, in the more general phenomenological representation of loss, σ is complex as well. In that case, $L^{(\nu)}$, $C^{(\nu)}$, and $\mathcal{N}_{\nu p}$ may all become complex.

If loss is represented by a complex s_{11} only, with σ pure real, the Q of the disk modes are unchanged from the Q of the bar modes as given by Equation 2.18,

$$Q_v = \frac{{}^R s_{11}}{{}^I s_{11}} \tag{2.41}$$

However, if one assumes a complex Poisson's ratio, a rather involved formula replaces Equation 2.41 as Equation 2.36 then requires K_v to be complex. When the real part of K_v is designated ${}^R K_v$, and when σ is separated into real and imaginary parts by

$$\sigma = {}^R\sigma + j\,{}^I\sigma \tag{2.42}$$

then in the case of complex σ, we find[52]

$$Q_v = \left[\frac{{}^I s_{11}}{{}^R s_{11}} + 2\,{}^I\sigma\left(\frac{1}{{}^R K_v^2 - 1 + {}^R\sigma^2} + \frac{{}^R\sigma}{1 + \frac{1}{2}\,{}^R\sigma^2}\right)\right]^{-1} \tag{2.43}$$

In this equation, terms in Q_v^{-2} and σ^4 are dropped. Equation 2.43 reduces to Equation 2.41 in the special case of zero ${}^I\sigma$.

2.2.2. Device Applications

Symmetric two- and three-electrode-pair ferroelectric ceramic disks have the same device applications as the corresponding bar combinations: tuned transformers, electrically tuned oscillators, FM discriminators, etc. In fact, disk configurations possess distinct advantages over bar configurations for electrically tuned oscillators and FM discriminators.

For instance, refer to the electrically tuned resonator equivalent circuit in Figure 2.5. This equivalent circuit may be applied equally well to bar or disk geometry resonators. However, the all-important ratio of the variable capacitance $C_0(3)/\mathcal{N}_{v3}^{\,2}$ to the fixed capacitance $C^{(v)}$ will typically be about three times as great for disks as for bars of the same material. Consequently the tuning range which one can achieve in the disk configuration will be about three times as great as in the bar configuration.

For the same reason, one can build FM discriminators from disk configuration resonators which have a linear input frequency-output voltage bandwidth about three times as wide as can be obtained from bars.

2.3. General Formulation of the Admittance Matrix Problem for an Arbitrarily Shaped Plate with Arbitrary Electrode Configuration

2.3.1. General Theory

In the two previous sections, formulas were presented giving the admittance matrices Y_{pq} of ferroelectric ceramic contour-extensional bar and disk resonators electroded in fairly special manners. We shall now derive a parallel result which is valid for an arbitrarily shaped

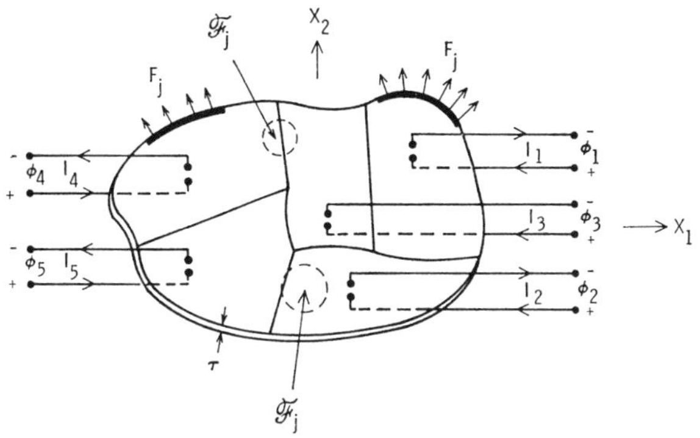

Figure 2.10. A piezoelectric plate covered by an arbitrary electrode configuration. The top surface and bottom surface electrode patterns are assumed to be the same.

contour-extensional thin-plate resonator fabricated from an arbitrary piezoelectric material and electroded with an arbitrary pattern.

Let us begin by considering the plate shown in Figure 2.10, which is completely covered on both sides by identical electrodes. The x_3-axis is defined to be perpendicular to the plate, as previously. It will be assumed that the plate thickness τ is small, so that electrical fringing is negligible and so that the stress tensor components T_{13}, T_{23}, and T_{33} are zero. Under these assumptions, there can be three nonzero stresses $T_1 = T_{11}$, $T_2 = T_{22}$, $T_6 = T_{12}$, and one nonzero electrical field E_3. Thus the matrix notation constituent relations of Equation 1.18 reduce to

$$\begin{aligned} S_\nu &= s^E_{\nu\mu} T_\mu + d_{3\nu} E_3 \\ D_3 &= d_{3\mu} T_\mu + \epsilon^T_{33} E_3 \end{aligned} \tag{2.44}$$

where μ and ν here range only over 1, 2, and 6.

The derivation that we present in this section is most easily carried out with strain and electric field as the independent variables. In order to effect this rearrangement of Equation 2.44, we define the planar stiffness matrix $\gamma^E_{\mu\lambda}$ by

$$s^E_{\nu\mu}\gamma^E_{\mu\lambda} = \delta_{\nu\lambda} \tag{2.45}$$

where again all subscripts range only over 1, 2, and 6. (Thus $\gamma^E_{\mu\nu}$ differs from $c^E_{\mu\lambda}$ of Equation 1.24 in being the inverse of the (1,2,6) $s^E_{\nu\mu}$ submatrix, not the entire 6 × 6 matrix.) Application of this definition permits us to write Equation 2.44 as

$$\begin{aligned} T_\nu &= \gamma^E_{\nu\mu}S_\mu - \tilde{e}_{3\nu}E_3 \\ D_3 &= \tilde{e}_{3\mu}S_\mu + \epsilon^{S_2}_{33}E_3 \end{aligned} \tag{2.46}$$

where

$$\begin{aligned} \tilde{e}_{3\nu} &= d_{3\mu}\gamma^E_{\mu\nu} \\ \epsilon^{S_2}_{33} &= \epsilon^T_{33} - d_{3\mu}d_{3\nu}\gamma^E_{\mu\nu} \end{aligned} \tag{2.47}$$

One can show, as in Equation 2.33, that $\epsilon^{S_2}_{33}$ is the dielectric constant evaluated with mechanical motion prevented in the plane of the plate but not perpendicular to the plate.

However, it is most convenient to work not with Equations 2.46 but with their tensor equivalent,

$$\begin{aligned} T_{ij} &= \gamma^E_{ijkl}S_{kl} - \tilde{e}_{3ij}E_3 \\ D_3 &= \tilde{e}_{3kl}S_{kl} + \epsilon^{S_2}_{33}E_3 \end{aligned} \tag{2.48}$$

Here i, j, k, l are understood to range over 1 and 2 only and are related to μ and ν by Equation 1.14. Consistency of Equation 2.46 with 2.48 requires the tensor and matrix γ and $\tilde{e}$ coefficients to be related by

$$\begin{aligned} \gamma^E_{ijkl} &= \gamma^E_{\nu\mu} \\ \tilde{e}_{3kl} &= \tilde{e}_{3\mu} \end{aligned} \tag{2.49}$$

in analogy to Equation 1.30.

This completes our rearrangement of the constituent relations, as Equation 2.48 is the optimum form for our present purposes. As in the previous sections, we will assume that all electroelastic variables contain a suppressed $e^{j\omega t}$ dependence. All coefficients may again be considered complex, in order to represent the most general phenomenological loss. In this section the coefficients, including the elastic ones, will be permitted to vary over the plate area. Density ρ will also be permitted to have x_1x_2-dependence, although τ will be assumed uniform, and all quantities will be assumed x_3-independent.

In the previous sections, only mechanical forces were allowed which acted at the extremity of the sample. For added generality, we shall here treat the case in which a force per unity area $\mathscr{F}_j$ is applied anywhere in the plane of the plate (see Figure 2.10). This force may represent either an external mechanical drive or a mechanical load. Its inclusion in our discussion here permits the results of our theory to be applied in transducer design.

Under these assumptions and conditions, the differential equation of motion for the plate becomes

$$-\rho\omega^2 u_j = T_{ij,i} + \mathscr{F}_j/\tau \tag{2.50}$$

where commas in this section represent covariant differentiation. Combining Equations 2.48 and 2.50 leads to a wave equation which is valid over the entire plate,

$$(\gamma^E_{ijkl} u_{k,l})_{,i} + \rho\omega^2 u_j = (\tilde{e}_{3ij} E_3)_{,i} - \mathscr{F}_j/\tau \tag{2.51}$$

If the plate is free, $\mathscr{F}_j$ is zero and the equation is homogeneous everywhere except at the electrode edges. Equation 2.51 clearly demonstrates the equivalence of the piezoelectric force localized at the electrode edges $-(\tilde{e}_{3ij}E_3)_{,i}$ to the mechanical driving term $\mathscr{F}_j/\tau$.

At the risk of laboring the point, we desire to emphasize this concept dealing with the nature of piezoelectric drive, which is probably the most significant aspect of the present chapter. Consider the thin plates shown in Figure 2.11 with voltages applied to the electrodes as illustrated. Intuitively, we would expect the piezoelectric effect of these applied voltages to be some sort of a body force density, uniform under

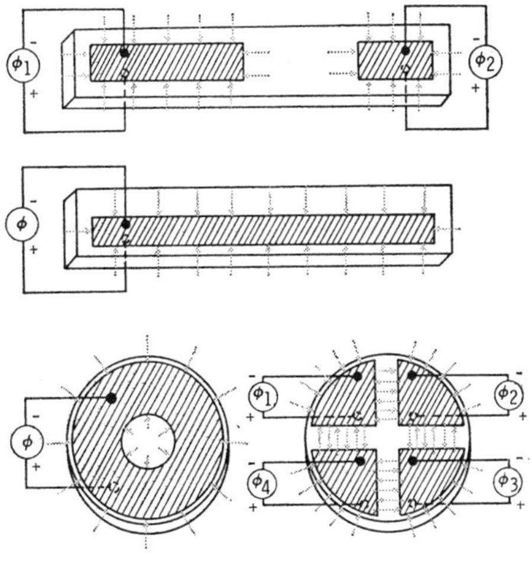

Figure 2.11. The piezoelectric force distribution.

a given electrode. However, the remarkable fact is that the piezoelectric force caused by these voltages is actually a line singularity acting only at the edge of each electrode region. No forces are present inside the electroded region itself. The grey vectors shown in Figure 2.11 thus indicate the locus at which this piezoelectric force acts.

To conclude the formulation of the problem we deal with in this section, let us give the mechanical boundary condition at the edge of the plate in question. If a force per unit length F_j is applied in an outward direction around the edge of the plate (see Figure 2.10) this boundary condition may be written

$$N_j(\gamma^E_{ijkl}u_{k,l} - \tilde{e}_{3ij}E_3) = F_i/\tau \tag{2.52}$$

Here, N_j is the outward pointing unit normal vector.

Equations 2.51 and 2.52 constitute an inhomogeneous boundary value problem for the displacement u_k. Analogous boundary value problems with scalar unknowns are referred to as Sturm-Liouville systems.[53,54] We shall now illustrate a method of extending the conventional Green's function approach to Sturm-Liouville problems so that it is applicable to the present vector differential equation system.

We shall have necessity to discuss points on the plate from two viewpoints: Let **r** represent a point at which drive (either mechanical or piezoelectric) is being applied, and let **R** represent a point at which resulting motion is being observed. We may thus refer to **r** as the source coordinate and **R** as the observer coordinate, in analogy with x' and x'' in Section 2.1.

It is possible to define vector transformations from the source to the observer system. For example, if we characterize vectors evaluated at **r** by lower case indices and vectors evaluated at **R** by upper case indices, then δ_{iI} represents the identity transformation or uniform translation from **r** to **R**:

$$X_i\delta_{iI} = X_I \tag{2.53}$$

where

$$X_i = X_i(\mathbf{r})$$
$$X_I = X_I(\mathbf{R}) \qquad (i, I = 1, 2) \tag{2.54}$$

We note that this transformation is independent of the points **r** and **R** at which it operates. A more general vector transformation could, of course, be defined to depend on both **r** and **R**.

The Green's function relevant to the problem at hand may be regarded as such a vector transformation depending on **r** and **R**. We shall represent it by $G(\mathbf{r} \mid \mathbf{R})_{iI}$, and define it to obey the differential equation

$$[\gamma^E_{ijkl} G(\mathbf{r} \mid \mathbf{R})_{kI,l}]_{,i} + \omega^2 \rho G(\mathbf{r} \mid \mathbf{R})_{jI} = -\delta(\mathbf{r} - \mathbf{R})\delta_{jI}/\tau \tag{2.55}$$

In Equation 2.55, the $\delta(\mathbf{r} - \mathbf{R})$ is the two-dimensional Dirac delta function.

Equation 2.55 does not completely specify the Green's function: Some stress-type boundary condition such as Equation 2.52 must be given as well. The most useful result will subsequently be obtained if we specify that boundary condition to be homogeneous:

$$N_i \gamma^E_{ijkl} G(\mathbf{r} \mid \mathbf{R})_{kI,l} = 0 \tag{2.56}$$

Having defined the Green's function by Equations 2.55 and 2.56 we now utilize it in the conventional way by writing an interaction integral between $u_j(\mathbf{r})$ and $G(\mathbf{r} \mid \mathbf{R})_{kI}$. In this case, this integral, which is analogous to Equation 2.11, is an $\mathbf{R}$-system vector,

$$J_I = \iint_A \{u_j(\mathbf{r})[\gamma^E_{ijkl} G(\mathbf{r} \mid \mathbf{R})_{kI,l}]_{,i} - [\gamma^E_{ijkl} u_{k,l}(\mathbf{r})]_{,i} G(\mathbf{r} \mid \mathbf{R})_{jI}\} \, dA \tag{2.57}$$

Here, A is the total surface area of the plate and integration is over $\mathbf{r}$. This integral has properties which are very useful in solving Equations 2.51 and 2.52. Application of the product rule for derivatives and of the divergence theorem enables us to convert this surface integral over the plate to a contour integral around the plate edge ∂A.

$$J_I = -\oint_{\partial A} \left(\frac{F_j}{\tau} + N_i \tilde{e}_{3ij} E_3\right) G(\mathbf{r} \mid \mathbf{R})_{jI} \, ds \tag{2.58}$$

The boundary conditions, Equations 2.52 and 2.56, have also been used in obtaining Equation 2.58.

On the other hand, substitution of the differential equations 2.51 and 2.55 into 2.57 yields

$$\begin{aligned} J_I = &-\iint_A u_j(\mathbf{r})\delta(\mathbf{r} - \mathbf{R})\delta_{jI} \, dA/\tau \\ &-\iint_A [(\tilde{e}_{3ij} E_3)_{,i} - \mathscr{F}_j/\tau] G(\mathbf{r} \mid \mathbf{R})_{jI} \, dA \end{aligned} \tag{2.59}$$

It is consequently possible to eliminate J_I from Equations 2.58 and

2.59, and thus to find

$$\iint_A u_j(\mathbf{r})\delta(\mathbf{r} - \mathbf{R})\delta_{jI}\, dA$$

$$= -\iint_A [\tau(\tilde{e}_{3ij}E_3)_{,i} - \mathscr{F}_j]G(\mathbf{r} \mid \mathbf{R})_{jI}\, dA + \oint_{\partial A} [\tau N_i \tilde{e}_{3ij}E_3 + F_j]G(\mathbf{r} \mid \mathbf{R})_{jI}\, ds \qquad (2.60)$$

If we take advantage of the definition of the Dirac delta, the left side of Equation 2.60 goes explicitly to $u_I(\mathbf{R})$, the particle displacement at the point of observation. The right side of Equation 2.60 becomes an integral of the source or driving terms weighted by Green's function. Consequently, what we must now do is to evaluate the Green's function, which is the only unknown in this formula for $u_I(\mathbf{R})$.

In order to do this, let us consider the homogeneous form of Equations 2.51 and 2.52 for u_j. This system will have nontrivial solutions only for certain discrete complex frequencies ω_ν. These frequencies are often called the eigenfrequencies of the system, and the corresponding displacement functions are called the eigenmodes:

$$\begin{aligned} (\gamma^E_{ijkl}u^{(\nu)}_{k,l})_{,i} + \rho\omega^2_\nu u^{(\nu)}_j &= 0 \\ N_j\gamma^E_{ijkl}u^{(\nu)}_{k,l} &= 0 \end{aligned} \qquad (2.61)$$

As the $u^{(\nu)}_j$ are obtained by setting E_3 equal to zero in Equation 2.51, they represent the short-circuit eigenmodes. Note that in the plane-stress, no-fringing approximation these eigenmodes are independent of the dielectric and piezoelectric properties of the plate.*

Since the $u^{(\nu)}_j$ are defined with $E_3 = 0$, the second of Equations 2.48 may be used to associate an electric displacement distribution with these eigenmodes:

$$D^{(\nu)}_3 = \tilde{e}_{3kl}u^{(\nu)}_{k,l} \qquad (2.62)$$

It is known that the eigenmodes of a system of equations such as Equation 2.61 are orthogonal when weighted by the material density ρ.[55] (If ω^2_ν is a degenerate eigenvalue, the usual diagonalization procedure will be necessary to make the eigenmodes corresponding to

* The solutions to Equations 2.61 describe the actual resonant behavior of a plate of thickness τ only for modes for which τ is a small fraction of an acoustic wavelength.

ω_ν orthogonal among themselves.) It will be convenient to normalize these eigenmodes according to the rule

$$\iint_A \rho u_i^{(\mu)} u_i^{(\nu)}\, dA = \frac{\delta_{\mu\nu}}{\tau} \tag{2.63}$$

We now postulate that $G(\mathbf{r} \mid \mathbf{R})_{iI}$ can be expressed as a Fourier-eigenmode series in the $u_i^{(\nu)}$:

$$G(\mathbf{r} \mid \mathbf{R})_{iI} = \sum_\nu B_I^{(\nu)}(\mathbf{R}) u_i^{(\nu)}(\mathbf{r}) \tag{2.64*}$$

The $B_I^{(\nu)}(\mathbf{R})$ are the Fourier coefficients to be determined. This solution satisfies the boundary condition Equation 2.56 as a term-by-term comparison with the second of Equations 2.61 shows. The $B_I^{(\nu)}$ may be selected so that Equation 2.64 satisfies Equation 2.55, the differential equation for $G(\mathbf{r} \mid \mathbf{R})_{iI}$ as well. In particular, if we substitute Equation 2.64 into Equation 2.55, multiply both sides of the result by $u_j^{(\mu)}(\mathbf{r})$, and apply the first of Equations 2.61 and 2.63, we find

$$G(\mathbf{r} \mid \mathbf{R})_{iI} = \sum_\nu \frac{u_i^{(\nu)}(\mathbf{r}) u_I^{(\nu)}(\mathbf{R})}{\omega_\nu^2 - \omega^2} \tag{2.65}$$

If Equation 2.65 for the Green's function is substituted into Equation 2.60, one obtains the following expression for the particle displacement at the point of observation:

$$u_I(\mathbf{R}) = -\iint_A [\tau(\tilde{e}_{3ij}E_3)_{,i} - \mathscr{F}_j] \sum_\nu \frac{u_j^{(\nu)}(\mathbf{r}) u_I^{(\nu)}(\mathbf{R})\, dA}{\omega_\nu^2 - \omega^2}$$
$$+ \oint_{\partial A} [\tau N_i \tilde{e}_{3ij} E_3 + F_j] \sum_\nu \frac{u_j^{(\nu)}(\mathbf{r}) u_I^{(\nu)}(\mathbf{R})\, ds}{\omega_\nu^2 - \omega^2} \tag{2.66}$$

This equation may be simplified considerably by applying the divergence theorem and the product derivative rule to the piezoelectric terms. It is also convenient hereafter to regard the forces on the edge of the plate F_j as a line singularity contribution to the force per unit area $\mathscr{F}_j$ acting on the plate surface. If we designate the total force per unit area

* This summation and all subsequent summations on the eigenmodes in this section are understood to range over the modes for which plate thickness is a small fraction of an acoustic wavelength. For ω low enough to satisfy this condition, Equation 2.64 has the proper form to satisfy Equations 2.55 and 2.56 for $G(\mathbf{r} \mid \mathbf{R})_{kI}$.

($\mathscr{F}_j$ and F_j) by $\mathscr{F}'_j$, we can obtain the result from Equation 2.66,

$$u_I(\mathbf{R}) = \sum_\nu u_I^{(\nu)}(\mathbf{R}) \frac{\left[\iint_A \tau \tilde{e}_{3kl} E_3 u_{k,l}^{(\nu)}(\mathbf{r})\, dA + \iint_A \mathscr{F}'_j u_j^{(\nu)}(\mathbf{r})\, dA\right]}{\omega_\nu^2 - \omega^2} \tag{2.67}$$

We have now succeeded in evaluating the particle displacement at an arbitrary point of observation $u_I(\mathbf{R})$ in terms of the properties of the plate and the driving forces and voltages. The next step is to obtain a similar expression for the current at an arbitrary electrode pair. To do this, let us consider a plate having N electrode pairs indexed by p and q. The current at electrode pair q is the time derivative of the electric displacement integrated over the area of electrode region q. In view of Equation 2.48, this becomes

$$I_q = j\omega \iint_{A_q} (\tilde{e}_{3ij} u_{i,j} + \epsilon_{33}^{S_2} E_3)\, dA \tag{2.68}$$

where A_q is the portion of the plate covered by electrode region q. Substitution of the particle displacement from Equation 2.67 then yields

$$I_q = j\omega \sum_\nu \left[\iint_{A_q} \tilde{e}_{3ij} u_{i,j}^{(\nu)}\, dA \cdot \frac{\left\{\iint_A \tau \tilde{e}_{3kl} E_3 u_{k,l}^{(\nu)}\, dA + \iint_A \mathscr{F}'_k u_k^{(\nu)}\, dA\right\}}{\omega_\nu^2 - \omega^2} \right] + j\omega \phi_q C_0(q) \tag{2.69}$$

where $\phi_q = \tau E_3$ is the voltage applied at electrode pair q, and where $C_0(q)$ is the capacitance at electrode pair q evaluated with motion prevented in the x_1x_2-plane:

$$C_0(q) = \frac{1}{\tau} \iint_{A_q} \epsilon_{33}^{S_2}\, dA \tag{2.70}$$

It is now quite simple to evaluate the electrical portion of the desired admittance matrix for the device shown in Figure 2.10. In particular, this matrix is given by the derivative of I_q with respect to the voltage applied at electrode pair p, ϕ_p. If we observe that differentiation of τE_3 by ϕ_p yields 1 at electrode region p and 0 elsewhere, we see from

Equation 2.69 that

$$\frac{\partial I_q}{\partial \phi_p} = Y_{qp} = j\omega \sum_{\nu} \frac{\iint_{A_q} D_3^{(\nu)}\,dA \iint_{A_p} D_3^{(\nu)}\,dA}{\omega_\nu^2 - \omega^2} + j\omega C_0(q)\delta_{pq} \tag{2.71}$$

Equation 2.62 for $D_3^{(\nu)}$ has also been used in obtaining this result. The electrical reciprocity of the configuration in question is readily apparent from this relation.

It is not quite so straightforward to obtain the mechanical and electromechanical components of the admittance matrix; some definitions must be introduced first. Let us assume that there are M independent force distributions acting on the sample (where M may be infinite). Let these distributions be indexed by t and w. Under this assumption, we can expand $\mathscr{F}_i'$ in the form

$$\mathscr{F}_i' = \sum_{t=1}^{M} \mathscr{F}_i^{(t)} \phi_t' \tag{2.72}$$

where the $\mathscr{F}_i^{(t)}$ represent the M distribution patterns and the ϕ_t' are weighting factors multiplying those distributions.

The electromechanical portion of the admittance matrix is obtained by differentiating I_q as given in Equation 2.69 with respect to ϕ_t' as defined by Equations 2.72:

$$\frac{\partial I_q}{\partial \phi_t'} = Y_{qt} = j\omega \sum_{\nu} \frac{\iint_{A_q} D_3^{(\nu)}\,dA \iint_{A} \mathscr{F}_k^{(t)} u_k^{(\nu)}\,dA}{\omega_\nu^2 - \omega^2} \tag{2.73}$$

The ϕ_t' weighting factors may be regarded as the mechanical voltage analogues in this system. Corresponding mechanical current analogues may be defined by*

$$I_w' = j\omega \iint_{A} \mathscr{F}_k^{(w)} u_k\,dA \tag{2.74}$$

* The product $\mathscr{F}_k^{(w)}\phi_w'$ has dimensions of force/area in this formulation. Similarly the product $\phi_w' I_w'$ has dimensions of power. However, the quantities $\mathscr{F}_k^{(w)}$, ϕ_w', and I_w' are individually only determined within a multiplicative constant by Equations 2.72 and 2.74. That constant is not necessarily dimensionless and may even depend on w; its choice is dictated by convenience. As there is no single choice which is universally most convenient, $\mathscr{F}_k^{(w)}$, ϕ_w', and I_w' may have w-dependent dimensionality in a given problem and different dimensions in different problems.

In the case of a point force application, $\mathscr{F}_k^{(w)}$ is a delta function, and this definition reduces to particle velocity, which is a conventional current analogue. In the more general case of distributed force application, this definition leads to a convenient electromechanical reciprocity, which is discussed later.

Using Equation 2.74, we may alternatively express the electromechanical portion of the admittance matrix as the derivative of I'_w with respect to ϕ_p. If we substitute Equation 2.67 for u_k into Equation 2.74, we obtain

$$\frac{\partial I'_w}{\partial \phi_p} = Y_{wp} = j\omega \sum_{\nu} \frac{\iint_{A_p} D_3^{(\nu)}\, dA \iint_{A} \mathscr{F}_k^{(w)} u_k^{(\nu)}\, dA}{\omega_\nu^2 - \omega^2} \tag{2.75}$$

Comparison of Equations 2.73 and 2.75 indicates that the electromechanical portion of the admittance matrix also possesses reciprocity. This reciprocity is a consequence of the definition of I'_w.

Finally, we may obtain the mechanical portion of the admittance by differentiating I'_w with respect to ϕ'_t,

$$\frac{\partial I'_w}{\partial \phi'_t} = Y_{wt} = j\omega \sum_{\nu} \frac{\iint_{A} u_i^{(\nu)} \mathscr{F}_i^{(t)}\, dA \iint_{A} u_k^{(\nu)} \mathscr{F}_k^{(w)}\, dA}{\omega_\nu^2 - \omega^2} \tag{2.76}$$

The piezoelectric plate shown in Figure 2.10 represents a distributed medium problem with an infinite set of resonances. As such, it cannot be represented for all ω by a finite number of lumped equivalent circuit elements. However, usually we are interested in operating a piezoelectric device near a resonance which is well separated from all other resonances. Under this circumstance, only one term of each admittance matrix series, Equations 2.71, 2.73, 2.75, and 2.76, will contribute significantly. If this term is designated as the νth, the admittance matrix of the plate shown in Figure 2.10 may be represented by the equivalent circuit given in Figure 2.12. The elements of this circuit are related to the material coefficients and geometrical parameters of the plate by

$$L^{(\nu)}C^{(\nu)} = \omega_\nu^{-2} \tag{2.77}$$

$$\mathscr{N}_{\nu p} = \frac{1}{\omega_\nu C^{(\nu)1/2}} \iint_{A} D_3^{(\nu)}\, dA \tag{2.78}$$

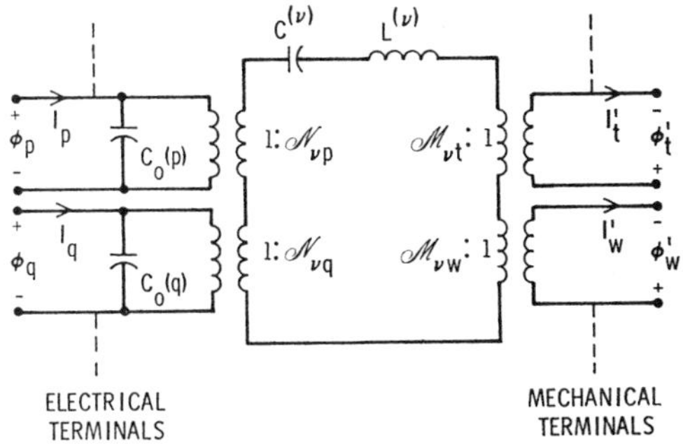

Figure 2.12. The equivalent circuit of an arbitrarily electroded piezoelectric plate near the νth resonance.

and

$$\mathscr{M}_{\nu t} = \frac{1}{\omega_\nu C^{(\nu)1/2}} \iint_A u_i^{(\nu)} \mathscr{F}_i^{(t)}\, dA \tag{2.79}$$

The nature of the transformer arrangement in this equivalent circuit is such that one of the following, $\mathscr{M}_{\nu t}$ $(t = 1, 2, \ldots, M)$, $\mathscr{N}_{\nu p}$ $(p = 1, 2, \ldots, N)$, $L^{(\nu)}$, or $C^{(\nu)}$, may be chosen arbitrarily. Once this has been done, the other circuit elements are all specified by Equations 2.77 to 2.79.

2.3.2. Specialization of the Mechanical Variable Definition to Point Force Application

In many cases of practical interest, external mechanical forces or loads are essentially applied to the plate at a point. This is the case, for example, if the plate is supported by the wires connecting the electrodes to external circuitry. Let us now examine the form taken by our definitions of mechanical terminal variables in this case.

A point force applied at $\mathbf{r}_0$ may always be expressed in a form compatible with Equation 2.72 by the equation

$$\mathscr{F}_i' = \sum_{t=1}^{2} \delta(\mathbf{r} - \mathbf{r}_0) a_i^{(t)} \phi_t' \tag{2.80}$$

where the $a_i^{(t)}$ are any two convenient orthogonal unit vectors, such as the unit vectors along the x_1- and x_2-axes. Thus, each point force can be represented by two independent force distribution patterns, which are given by

$$\mathscr{F}_i^{(t)} = \delta(\mathbf{r} - \mathbf{r}_0) a_i^{(t)} \qquad t = 1, 2 \tag{2.81}$$

and which correspond to the two degrees of freedom of a point in a plane. Mechanical current analogues are obtained by substituting Equation 2.81 into 2.74:

$$I'_w = j\omega a_k^{(w)} u_k(\mathbf{r}_0) \tag{2.82}$$

The equivalent transformer turns ratios for the mode ν equivalent circuit are obtained by substituting Equation 2.81 into 2.79:

$$\mathscr{M}_{\nu t} = u_i^{(\nu)}(\mathbf{r}_0) a_i^{(t)} / \omega_\nu C^{(\nu)^{1/2}} \tag{2.83}$$

If the point force acting at $\mathbf{r}_0$ represents an independent driving source, Equations 2.80 to 2.83 completely describe that source in a manner consistent with the previously developed formulation. It is represented by two mechanical terminal pairs having equivalent voltages specified by Equation 2.80 and equivalent currents specified by Equation 2.82.

However, usually a point force is of a loading rather than a driving nature. For example, we may have a mechanical load at $\mathbf{r}_0$ representable by a mechanical admittance tensor y_{kl} where

$$j\omega u_k(\mathbf{r}_0) = y_{kl} \sum_{t=1}^{2} a_l^{(t)} \phi'_t \tag{2.84}$$

In Equation 2.84, the y_{kl} relates the particle velocity at $\mathbf{r}_0$, $j\omega u_k(\mathbf{r}_0)$ to the total force applied by the load at $\mathbf{r}_0$, $\Sigma a_l^{(t)} \phi'_t$.

It is desirable to determine the effect of the mechanical admittance y_{kl} on the equivalent circuit of Figure 2.12. This effect may be evaluated by multiplying Equation 2.84 by $a_k^{(w)}$ in order to obtain I'_w as specified by Equation 2.82:

$$I'_w = a_k^{(w)} y_{kl} \sum_{t=1}^{2} a_l^{(t)} \phi'_t \tag{2.85}$$

Differentiating Equation 2.85 with respect to ϕ'_t then gives the admittance matrix of the equivalent two-port load,

$$(Y_{wt})_{\text{eq}} = a_k^{(w)} y_{kl} a_l^{(t)} \qquad w,t = 1,2 \tag{2.86}$$

This equivalent load may be connected to the two mechanical terminal pairs associated with $\mathbf{r}_0$ to incorporate the effects of y_{kl} into the mode ν equivalent circuit (see Figure 2.13).

2.3.3. Specialization of the Mechanical Variable Definitions to Distributed Force Application

If a distributed force is applied to the plate instead of point forces, the approach of Section 2.3.2 is obviously useless. For continuous

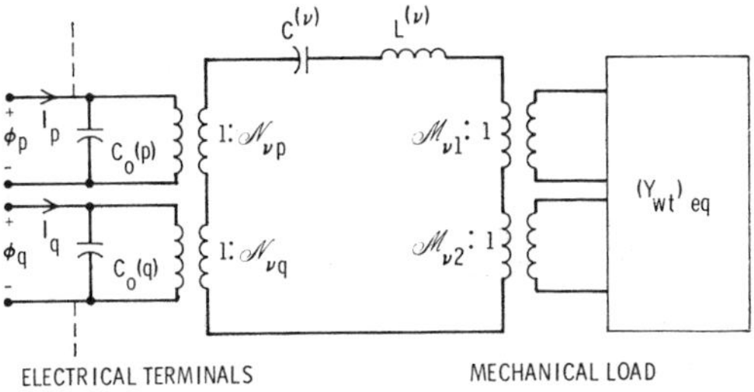

Figure 2.13. The effect of a mechanical point load on the mode ν equivalent circuit of a piezoelectric plate.

force distributions, one possibility here is to expand $\mathcal{F}'_k$ in terms of the mechanical eigenmodes:

$$\mathcal{F}'_k = \sum_t \tau \rho u_k^{(t)} \phi'_t \tag{2.87}$$

Other definitions of the mechanical analogues may be more efficient in cases of specific force distribution. However, Equation 2.87 is applicable to an arbitrary force distribution provided the usual assumptions regarding the completeness of the $u_k^{(t)}$ are valid.

There are now strictly speaking an infinite number of independent force distribution patterns $\mathcal{F}_k^{(t)}$ (see Equation 2.72):

$$\mathcal{F}_k^{(t)} = \tau \rho u_k^{(t)} \tag{2.88}$$

However, at frequencies low enough to make our plane-stress approximation valid, we need only consider the finite number of force distribution patterns which are not excluded by the footnote to Equation 2.64. Equivalent mechanical currents arising from this treatment may be obtained by substituting Equations 2.67, 2.87, and 2.88 into 2.74:

$$I'_t = \frac{j\omega\left(\iint\limits_A \tau \tilde{e}_{3ij} E_3 u_{i,j}^{(t)}\, dA + \phi'_t\right)}{\omega_t^2 - \omega^2} \tag{2.89}$$

The orthogonality relation, Equation 2.63, is used several times in deriving this result. It may be observed that if the mechanical terminal variables ϕ'_t and I'_t are defined in the manner of Equation 2.87, there is no transfer admittance between the various mechanical ports:

$$Y_{tw} = \frac{j\omega\delta_{tw}}{\omega_t^2 - \omega^2} \tag{2.90}$$

Alternatively, substitution of Equation 2.88 for $\mathscr{F}_k^{(t)}$ into Equation 2.79 for $\mathscr{M}_{vt}$, the mechanical transformer turns ratio, yields

$$\mathscr{M}_{vt} = \frac{\delta_{vt}}{\omega_v C^{(v)^{1/2}}} \tag{2.91}$$

Consequently, while a large number of mechanical terminals are required by expansion Equation 2.87, only one of these terminals (the νth) will have a nonzero transformer coupling ratio to the external mechanical forces in the equivalent circuit for operation near mode ν.

Equations 2.87 to 2.91 are sufficient to account for an independent distributed mechanical drive. However, as in the case of the point force, the characterization of a distributed mechanical load is somewhat more difficult.

Let us assume a distributed mechanical load is present which is representable by the admittance tensor η_{kl}, so that

$$j\omega u_k = \eta_{kl}\mathscr{F}'_l = \eta_{kl} \sum_t \tau\rho u_l^{(t)} \phi'_t \tag{2.92}$$

In other words, let the particle velocity be locally related to the applied force per unit area by η_{kl}. We now desire to relate this η_{kl} to some equivalent mechanical load. To do this, we multiply Equation 2.92 by $\tau\rho u_k^{(w)}$ and integrate over A, thus obtaining I'_w as defined in Equation 2.74:

$$I'_w = \iint_A \tau\rho u_k^{(w)} \sum_t \tau\rho\eta_{kl} u_l^{(t)} \phi'_t \, dA \tag{2.93}$$

Differentiation with respect to ϕ'_t yields the admittance matrix of a multiterminal mechanical load representing the effects of η_{kl} on the plate.

$$(Y_{wt})_{\text{eq}} = \iint_A \tau^2 \rho^2 u_k^{(w)} \eta_{kl} u_l^{(t)} \, dA \tag{2.94}$$

Connecting this multiterminal load to the equivalent circuit of Figure 2.12 then characterizes pictorially the effect of η_{kl} on the plate near mode ν (see Figure 2.14). As Figure 2.14 shows, only the νth terminal pair of this load is actually coupled into the mechanical system in the vicinity of the νth isolated resonance. By virtue of Equation 2.91, all the other terminals of this equivalent mechanical load are connected to the "1" side of a 0:1 transformer and thus in effect are open circuited. Consequently, the effect of this mechanical load is to place an impedance of value $Z_{\nu\nu}$ across the νth mechanical terminal pair

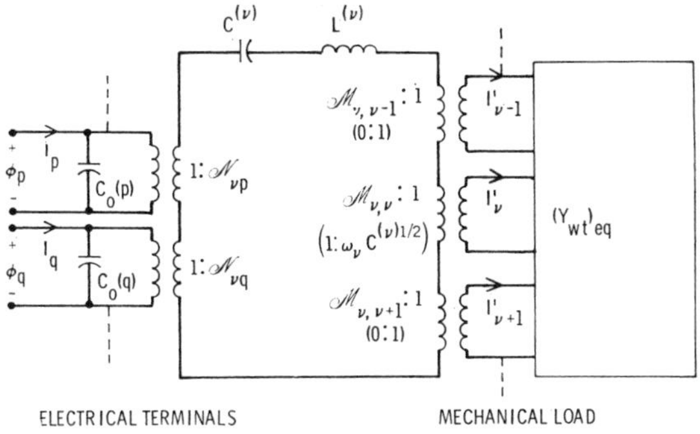

Figure 2.14. The effect of a mechanical distributed load on the equivalent circuit of a piezoelectric plate.

of the circuit of Figure 2.14, where*

$$Z_{wt} = [(Y_{wt})_{\text{eq}}]^{-1} \tag{2.95}$$

As the "0" side of a 0:1 transformer is by definition a short circuit, all the mechanical transformers in Figure 2.14 except the νth look like short circuits when considered from the side opposite the mechanical load and thus may be omitted. These operations are illustrated in Figure 2.15.

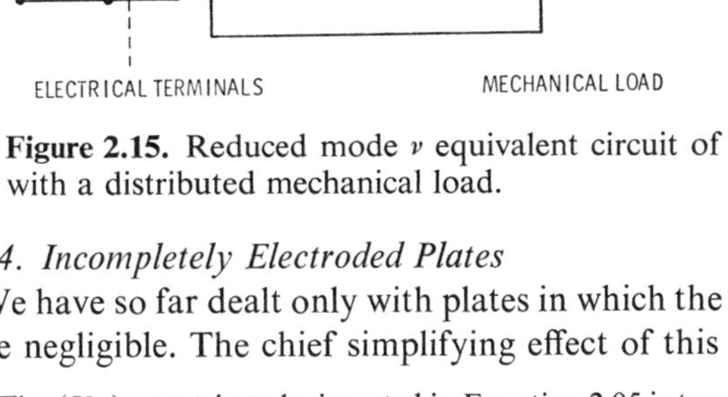

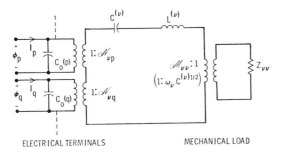

Figure 2.15. Reduced mode ν equivalent circuit of a piezoelectric plate with a distributed mechanical load.

2.3.4. *Incompletely Electroded Plates*

We have so far dealt only with plates in which the unelectroded areas were negligible. The chief simplifying effect of this assumption is that

* The $(Y_{wt})_{\text{eq}}$ matrix to be inverted in Equation 2.95 is truncated according to the same rule as that set forth in the footnote to Equation 2.64 for the truncation of series: Matrix elements corresponding to modes which violate the plane-stress assumption are excluded.

short circuiting all the electrodes then guarantees that E_3 will be everywhere zero.

However, there is a fairly easy technique for extending the complete electroding results to the partial electroding case. In particular, let us imagine the unelectroded areas to be completely covered by infinitesimal, nontouching, open-circuited electrode pairs. Excepting fringing effects between the infinitesimal electrode pairs, this imaginary arrangement is physically equivalent to the unelectroded situation. However, as the piezoelectrically induced voltages on adjacent infinitesimal electrode pairs will differ only by infinitesimal amounts, fringing effects between these adjacent electrode pairs in the imaginary arrangement will be on the order of second-order differentials and thus negligible. Consequently, near mode ν the effects of the unelectroded areas may be well represented in the equivalent circuit by an infinite set of electrical terminals, each open circuited except for infinitesimal capacitors representing clamped capacitance effects of the infinitesimal electrode pairs (see Figure 2.16a). We shall index the terminals of this infinite set by r.

Upon reflection through the associated transformers (Figure 2.16b), the infinitesimal capacitors representing clamped dielectric effects will individually present infinitesimal impedances of the form

$$Z_r = \frac{\mathcal{N}_{\nu r}^2}{j\omega C_0(r)} \tag{2.96}$$

where, according to Equations 2.70 and 2.78,

$$C_0(r) = \epsilon_{33}^{S_2} A_r/\tau \tag{2.97}$$

$$\mathcal{N}_{\nu r} = \frac{D_3^{(\nu)} A_r}{\omega_\nu C^{(\nu)^{1/2}}} \tag{2.98}$$

It should be observed that the area elements A_r are infinitesimal.

Upon summing these infinitesimal impedances and converting the summation to an integral, we obtain an overall equivalent capacitance for the unelectroded area (Figure 2.16c).

$$C_{\text{eq}} = \frac{\omega_\nu^2 C^{(\nu)}}{\tau} \left[\iint_{A_u} \frac{(D_3^{(\nu)})^2}{\epsilon_{33}^{S_2}} \, dA \right]^{-1} \tag{2.99}$$

Here A_u is the total unelectroded area. This equivalent capacitance enters the equivalent circuit in series with the motional capacitance of mode ν, $C^{(\nu)}$ (Figure 2.16c). Consequently unelectroded areas have the effect of raising resonant frequencies.

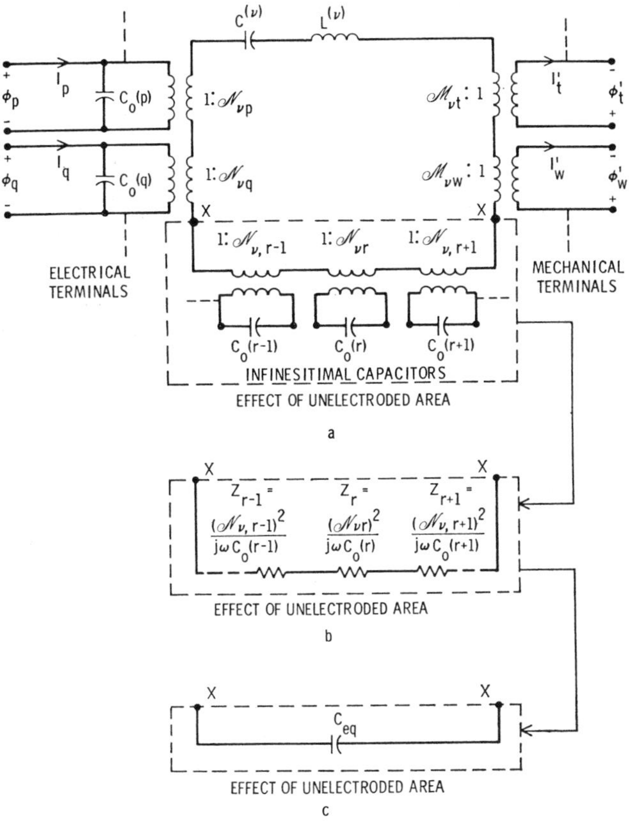

Figure 2.16. The effect of unelectroded areas on the equivalent circuit of a piezoelectric plate.

It is interesting to compare C_{eq} of Equation 2.99 with the equivalent capacitance in series with $C^{(\nu)}$ which represents the effect of an open-circuited, continuous electrode covering A_u. From Equations 2.70 and 2.78, we find

$$C'_{\text{eq}} = \frac{\omega_\nu^2 C^{(\nu)}}{\tau} \frac{\displaystyle\iint_{A_u} \epsilon_{33}^{S_2}\, dA}{\left[\displaystyle\iint_{A_u} D_3^{(\nu)}\, dA\right]^2} \tag{2.100}$$

Consequently we see that open-circuited electrodes also raise the resonant frequencies, although not by so much as the same area unelectroded.

2.3.5. *Example 1: A Bonded Down, Symmetrically Electroded Ferroelectric Ceramic Disk*

As a specific example of the techniques presented in this section, let us discuss the symmetrically electroded, axially poled ferroelectric ceramic disk shown in Figure 2.17. This disk is identical to the one considered in Section 2.2, except that there the major faces were free. In the present case, we shall assume one face of the disk is bonded to a substrate by an idealized bond representable by an admittance tensor (see Equation 2.92)

$$\eta_{kl} = \delta_{kl}\eta \tag{2.101}$$

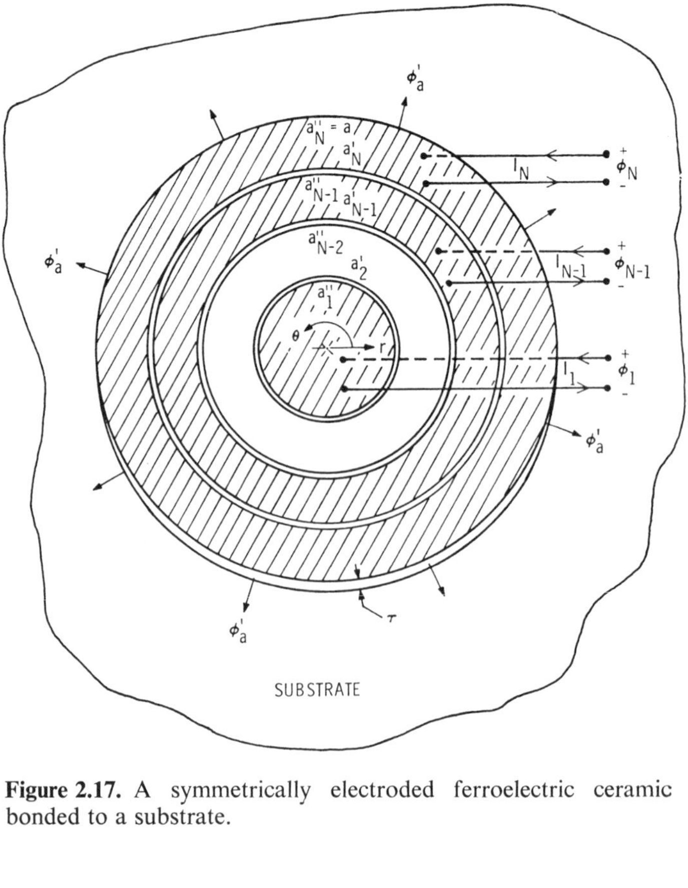

Figure 2.17. A symmetrically electroded ferroelectric ceramic disk bonded to a substrate.

In addition to the external mechanical forces associated with the bond, let us permit a uniform radial force to act around the edge of the disk as in Section 2.2. We shall again regard the edge of the disk as a mechanical terminal and shall again designate it by the subscript a (see Figure 2.17).

The arbitrary multiplicative constant discussed in the footnote to Equation 2.74 may be selected so that the mechanical analogue of current at this terminal is particle velocity,

$$I'_a = j\omega u_r(a) \tag{2.102}$$

The independent force distribution $\mathscr{F}_k^{(a)}$ associated with this mechanical terminal is to be applied uniformly around the edge of the disk in a radial direction. As the arbitrary multiplicative factor for this terminal has already been selected, consistency of Equations 2.74 and 2.102 determines $\mathscr{F}_k^{(a)}$ completely:

$$\mathscr{F}_k^{(a)} = \frac{\delta_{rk}\delta(r - a)}{2\pi a} \tag{2.103}$$

For simplicity, we shall assume η, ρ, and the elastic coefficients are uniform. The piezoelectric coefficients $\tilde{e}_{3ij}$ will be permitted to vary from electrode region to region but will be assumed to be uniform under a given electrode. It is known[43] that the eigenmode solutions to Equations 2.61 for this disk, normalized according to Equation 2.63, are

$$\begin{aligned} u_r^{(\nu)} &= \frac{J_1(rK_\nu/a)K_\nu}{aJ_1(K_\nu)\sqrt{\tau\rho\pi(K_\nu^2 - 1 + \sigma^2)}} \\ \omega_\nu^2 &= \frac{\gamma_{11}^E K_\nu^2}{\rho a^2} \end{aligned} \tag{2.104}$$

where σ is Poisson's ratio evaluated at constant E, $\gamma_{12}^E/\gamma_{11}^E$, and the K_ν are the roots of Equation 2.36

$$(1 - \sigma)J_1(K) = KJ_0(K) \tag{2.105}$$

Solutions of Equation 2.105 are given in Table 2.1.

We shall now illustrate the derivation of the equivalent circuit for this configuration. Recall that one of the quantities appearing on the left of Equations 2.77 to 2.79 may be selected arbitrarily. We choose to set

$$\mathscr{M}_{\nu a} = 1 \tag{2.106}$$

Equation 2.79 then specifies $C^{(\nu)}$,

$$C^{(\nu)} = [\pi\gamma_{11}^{E}(K_\nu^2 - 1 + \sigma^2)\tau]^{-1} \tag{2.107}$$

In evaluating this quantity, Equation 2.104 for $u_i^{(\nu)}$ and Equation 2.103 for $\mathscr{F}_i^{(a)}$ have been substituted. The equivalent motional inductance is then specified by Equations 2.77 and 2.104,

$$L^{(\nu)} = \pi a^2(K_\nu^2 - 1 + \sigma^2)\rho\tau K_\nu^{-2} \tag{2.108}$$

while Equations 2.62, 2.78, 2.104, and 2.107 serve to determine the $\mathscr{N}_{\nu p}$,

$$\mathscr{N}_{\nu p} = 2\pi a\tilde{e}_{311}(p)B_{\nu p} \tag{2.109}$$

In evaluating $\mathscr{N}_{\nu p}$, use is made of the fact that the only independent component of $\tilde{e}_{3ij}$ is $\tilde{e}_{311}$ for a ferroelectric ceramic. The symbol $B_{\nu p}$ appearing in Equation 2.109 is defined in Equation 2.34, and $\tilde{e}_{311}(p)$ represents $\tilde{e}_{311}$ evaluated at electrode region p.

The equivalent circuit quantities represented by Equations 2.102 to 2.109 agree with those of Section 2.2 for a ferroelectric ceramic disk mechanically free except at the circumference.

However, the methods of Section 2.3.3 are ideal for extending these results to the present case of a bonded down disk. Let us consider the equivalent admittance matrix of the bond as specified by Equation 2.94. Substituting Equation 2.101 into Equation 2.94, factoring $\rho\tau\eta$ outside the integral, and applying orthogonality relation Equation 2.63, we have

$$(Y_{wt})_{\mathrm{eq}} = \rho\tau\eta\delta_{wt} \tag{2.110}$$

This matrix may be readily inverted to give

$$Z_{wt} = \delta_{wt}/(\rho\tau\eta) \tag{2.111}$$

Consequently, the bond may be represented in the mode ν equivalent circuit by inserting an impedance of value

$$Z = Z_{\nu\nu}\mathscr{M}_{\nu\nu}^2 = \mathscr{M}_{\nu\nu}^2/\rho\tau\eta \tag{2.112}$$

in series with the motional capacitance $C^{(\nu)}$ (see Figure 2.18). Application of Equation 2.91 for $\mathscr{M}_{\nu\nu}$ then yields

$$Z = (\omega_\nu^2 C^{(\nu)}\rho\tau\eta)^{-1} \tag{2.113}$$

This impedance is equivalent to a complex capacitance of value

$$C = -j\omega_\nu C^{(\nu)}\rho\tau\eta \tag{2.114}$$

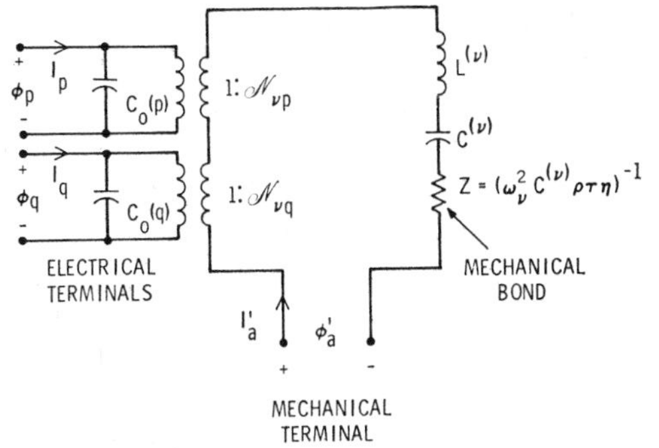

Figure 2.18. The mode ν equivalent circuit of a bonded down, symmetrically electroded ferroelectric ceramic disk.

so that the resultant capacitance of C_ν and C is

$$C_{\text{res}} = \frac{C^{(\nu)}}{1 + j/(\omega_\nu \rho\tau\eta)} \tag{2.115}$$

If we assume σ is pure real, the quality factor of the νth symmetric contour-extensional resonance Q_ν of a free disk ($\eta = 0$) is just the reciprocal loss tangent of $C^{(\nu)}$, as $L^{(\nu)}$ is ideal (see Equation 2.108 or the discussion at the end of Section 2.2). Moreover, according to Equation 2.107, the loss tangents of $C^{(\nu)}$ and γ_{11}^E are the same, so that

$$Q_\nu = \frac{{}^R\gamma_{11}^E}{{}^I\gamma_{11}^E} \tag{2.116}$$

for a free disk.

However, if the bonding admittance η is real (i.e., dissipative), this quality factor as obtained from Equation 2.115 will be reduced to

$$Q_\nu \doteq \frac{{}^R\gamma_{11}^E}{{}^I\gamma_{11}^E + (a/\tau)\cdot(\eta K_\nu)^{-1}\cdot\sqrt{{}^R\gamma_{11}^E/\rho}} \tag{2.117}$$

for the case of the bonded disk. Two interesting trends are apparent from Equation 2.117. It may be seen that the Q-reducing effects of the bond decrease with increasing resonance number and with increasing thickness to diameter ratio.

Experimental results. Equation 2.117 has been used to determine the admittance η characterizing RTV Silastic* bonds between a disk and substrate. This has been done by measuring the Q's of the first four

* Dow Corning trade name for a silicone-rubber adhesive.

resonances of symmetrically electroded ferroelectric ceramic disks, first when mechanically free, and second when bonded to a substrate with Silastic. The disks used were fabricated from the PZT 65/35 lead-zirconate-titanate composition described by Table 1.6.* Dimensions of the disks used here are 4.0 mils for τ and 62.5 mils for a.

Values for ${}^R\gamma_{11}^E$ may be obtained directly from Table 1.6 and Equation 2.45 with satisfactory precision. While we can, in principle, find ${}^I\gamma_{11}^E$ in the same way, it is more accurate to substitute ${}^R\gamma_{11}^E$, and the measured free Q_ν into Equation 2.117 with $\eta = 0$, and then to solve for ${}^I\gamma_{11}^E$. Then in the bonded case, all quantities appearing in Equation 2.117 except η are known. Measured values for Q_ν (free and bonded) for the first four disk modes are given in Table 2.2, along with the subsequently obtained values for η.

Table 2.2. Free and Bonded Disk Quality Factors and Corresponding Silastic Admittance

ν	Free Q_ν	Bonded Q_ν	η(MKS units)
1	140	40	0.109
2	128	59	0.081
3	116	54	0.047
4	98	52	0.038

These results, excluding the free Q_ν, were found to vary by as much as 30 percent as amplitude of excitation was varied. The free Q_ν, on the other hand, was essentially amplitude independent over the range studied. It may consequently be deduced that nonlinear loss effects were present in the Silastic but not in the ceramic.

Table 2.2 indicates that some frequency dependence in the free Q_ν, ${}^R\gamma_{11}^E/{}^I\gamma_{11}^E$, is present and that the Silastic admittance is fairly strongly frequency dependent.

It should probably be pointed out that these nonlinear and frequency-dependent characteristics of Silastic make it impossible to predict the Q-reducing effects of a Silastic bond with precision greater than 30 percent. (Perhaps other bonding materials will be found which are simpler to characterize.) However, for disks with only the inner third of the surface area bonded, the Q-reducing effects are on the order of 20 percent or less as contrasted with a maximum of 70 percent reported in Table 2.2 for a sample with the entire face uniformly bonded. Consequently, in the partial bonding case, which is actually the more

* Note that the coefficients given in Table 1.6 indeed yield σ pure real for PZT 65/35.

attractive from the practical viewpoint of attaining high Q's, it is possible to predict the resultant Q to within 30 percent of 20 percent; that is, to 6 percent.

2.3.6. *Example 2: An Asymmetrically Electroded, Free Ferroelectric Ceramic Disk*

Recently Onoe and Kurachi have determined the equivalent circuit of a one-electrode-pair axially poled, mechanically free ferroelectric ceramic disk for cases in which the electrode shape is θ-dependent.[56]

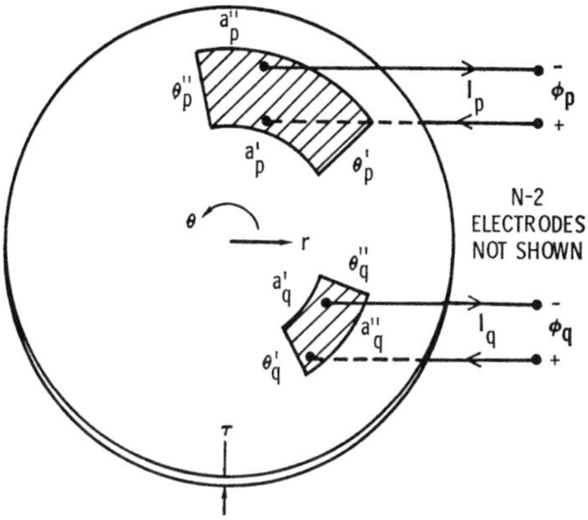

Figure 2.19. The *N*-electrode-pair asymmetric disk. It is assumed the bottom face is identically electroded.

Let us now apply the results of the present section to the same problem but relax the restriction that only one electrode pair be present. In particular, we shall consider a ferroelectric ceramic disk having N electrode regions, the pth bounded by $\theta'_p < \theta < \theta''_p$, $a'_p < r < a''_p$ (see Figure 2.19). The linear terminal behavior of this disk is completely specified by the electrode admittance matrix of Equation 2.71.

If we assume that $\tilde{e}_{3ij}$ does not vary under an individual electrode, the two integrals appearing in Equation 2.71 may be more conveniently evaluated in an alternative form,

$$\iint_{A_p} D_3^{(v)}\, dA = \oint_{\partial A_p} \tilde{e}_{3ij} u_i^{(v)} N_j^{(p)}\, ds \tag{2.118}$$

where ∂A_p is the edge of electrode region p and $N_j^{(p)}$ is the outward pointing unit vector normal to ∂A_p. Equation 2.62 and an integration by parts are required to establish this identity.

It should be noted that the summation of Equation 2.71 extends over all sufficiently low contour extensional modes of the disk. This includes modes lacking axial symmetry as well as the axially symmetric modes utilized in the previous example. The complete set of modes composes a two-dimensional series, so that it will be convenient to replace the single modal index ν with a double index (m,n) in the following discussion. In the double index convention, the first index is related to radial dependence and the second index is related to angular dependence. With this convention Equation 2.71 becomes

$$Y_{pq} = j\omega \sum_{m=1}^{\infty} \sum_{n=0}^{\infty} \frac{\oint_{\partial A_p} \tilde{e}_{3ij} u_i^{(m,n)} N_j^{(p)}\, ds \oint_{\partial A_q} \tilde{e}_{3kl} u_k^{(m,n)} N_l^{(q)}\, ds}{\omega_{m,n}^2 - \omega^2} + j\omega C_0(q)\delta_{pq} \tag{2.119}$$

The modes having zero angular dependence ($n = 0$) are specified by Equation 2.104 with

$$u_\theta^{(m,n)} = 0 \tag{2.120}$$

These displacement functions and the form of $\tilde{e}_{3ij}$ relevant to a ferroelectric ceramic lead to the result

$$\oint_{\partial A_p} \tilde{e}_{3ij} u_i^{(m,0)} N_j^{(p)}\, ds = \frac{K_{m,0}\tilde{e}_{311}(p) B_{mp}(\theta_p'' - \theta_p')}{\sqrt{\pi\rho\tau(K_{m,0}^2 - 1 + \sigma^2)}} \tag{2.121}$$

The $K_{m,0}$ appearing here are the K_ν of Equation 2.104.

The modes having nonzero angular dependence are considerably more complicated. They are given by[57]

$$u_r^{\binom{m,n,c}{m,n,s}} = \left[A_{m,n} \frac{dJ_n(K_{m,n}r/a)}{dr} + \frac{nB_{m,n}}{r} J_n(K_{m,n}\Theta r/a)\right] \cdot \frac{1}{\sqrt{\rho\tau}} \begin{pmatrix} \cos n\theta \\ -\sin n\theta \end{pmatrix}$$

$$u_\theta^{\binom{m,n,c}{m,n,s}} = -\left[\frac{nA_{m,n}}{r} J_n(K_{m,n}r/a) + B_{m,n} \frac{dJ_n(K_{m,n}\Theta r/a)}{dr}\right] \cdot \frac{1}{\sqrt{\rho\tau}} \begin{pmatrix} \sin n\theta \\ \cos n\theta \end{pmatrix} \qquad (m, n = 1, 2, \ldots) \tag{2.122}$$

The derivation of this result has been discussed by Onoe[58] and Love.[59] In Equation 2.122, the Θ is the ratio of dilational to shear acoustic velocity,

$$\Theta = \left(\frac{2}{1 - \sigma}\right)^{1/2} \tag{2.123}$$

The third index, c or s, in Equation 2.122 indicates whether the radial displacement component has sine or cosine angular dependence. These c and s modes are frequency degenerate but orthogonal. The summation of Equation 2.119 is understood to extend over both c and s modes. Here $A_{m,n}$ and $B_{m,n}$ are normalization coefficients which are discussed and tabulated in Reference 57. The $K_{m,n}$ are so-called eigenvalues of the problem and are analogous to the $K_{m,0}$ of the symmetric modes. These $K_{m,n}$ were first computed by Onoe,[58] although a more extensive listing of their values is available in Reference 57. The $K_{m,n}$ are related to the resonant frequency of mode (m,n) by

$$\omega_{m,n}^2 = \frac{\gamma_{11}^E K_{m,n}^2}{\rho a^2} \tag{2.124}$$

Substitution of Equation 2.122 into the integrals of Equation 2.119 yields

$$\int_{\partial A_p} \tilde{e}_{3ij} u_i^{\binom{m,n,c}{m,n,s}} N_j^{(p)}\, ds = \frac{\tilde{e}_{311}(p)}{\sqrt{\rho\tau}} \begin{pmatrix} \sin n\theta_p' - \sin n\theta_p'' \\ \cos n\theta_p' - \cos n\theta_p'' \end{pmatrix} \frac{A_{m,n} K_{m,n}^2}{na^2} \int_{a_p'}^{a_p''} r J_n(K_{m,n} r/a)\, dr \tag{2.125}$$

It should be observed that the $B_{m,n}$ coefficients of Equation 2.122 drop out of Equation 2.125. These coefficients are related to shear strains, which are piezoinactive for an axially poled ferroelectric ceramic. On the other hand, the $A_{m,n}$ coefficients are related to dilational strains, which are not piezoinactive.

If Equations 2.121 and 2.125 are substituted into Equation 2.119, a bit of algebra enables us to obtain the following formula for Y_{pq}:

$$Y_{pq} = j\omega C_0(q)\delta_{pq} + j\omega \tilde{e}_{311}(p)\tilde{e}_{311}(q) \cdot \Bigg\{ \sum_{\substack{m \\ m\neq 0}} \frac{K_{m,0}^2 B_{mp} B_{mq} (\theta_p'' - \theta_p')(\theta_q'' - \theta_q')}{\pi\rho\tau(K_{m,0}^2 - 1 + \sigma^2)(\omega_{m,0}^2 - \omega^2)} + \sum_{\substack{m \\ m\neq 0}} \sum_{\substack{n \\ n\neq 0}} \Bigg[\frac{4\cos\frac{1}{2}n(\theta_p'' - \theta_q'' + \theta_p' - \theta_q') \times \sin\frac{1}{2}n(\theta_p'' - \theta_p') \sin\frac{1}{2}n(\theta_q'' - \theta_q')}{\omega_{m,n}^2 - \omega^2} \cdot \frac{A_{m,n}^2}{n^2} I(m,n,p) I(m,n,q) \Bigg] \Bigg\} \tag{2.126}$$

In Equation 2.126, $I(m,n,p)$ represents the integral

$$I(m,n,p) = \int_{a_p'K_{m,n}/a}^{a_p''K_{m,n}/a} tJ_n(t)\, dt \tag{2.127}$$

Values for this integral have been tabulated.[60]

Equation 2.126 has been subjected to rather extensive experimental verification. These measurements, which have been published elsewhere,[61] are in quite good agreement with the present theory. As Equation 2.126 depends very intimately and in a complicated way on most of the basic concepts and techniques which are presented in this chapter, this agreement is regarded as constituting a sound verification of the work reported here.

2.3.7. Device Applications

There are a number of practical reasons for preferring the more complicated device configurations described in this section to those treated in Sections 2.1 and 2.2. For example, we may desire to use two- or three-electrode-pair symmetric disks for tuned transformers, electrically tuned oscillators, or FM discriminators, as described in Sections 2.1 and 2.2. In this case, the increased mechanical strength resulting from bonding the disk to a substrate as considered in Section 2.3.5 provides an obvious superiority. This is especially true when we reflect that many of the primary advantages of piezoelectric circuitry arise from its resistance to radiation, for example, in space or weapons systems applications: In these applications, components would require considerable structural integrity to withstand strong mechanical shock or acceleration.

However, in addition to these considerations, we can use the more complicated asymmetrically electroded disk for jobs the symmetric disk cannot perform. For example, consider the mechanically free ferroelectric ceramic disk with seven equal peripheral electrode regions and one central electrode region shown in Figure 2.20. The equivalent circuit for this configuration, according to the theory presented in this section, is shown in Figure 2.21 (see Figure 2.12 in the mechanically free case, for instance). Let the voltage inputs to the seven outer electrode pairs all be at the same frequency and close to one of the axisymmetric ($n = 0$ in Equation 2.119) resonances. Also, let the central electrode pair be terminated by a resistance. Then the voltage developed across the terminating resistance will be proportional to the phasor

Figure 2.20. A ferroelectric ceramic disk with seven equal peripheral electrode regions.

sum of the seven input voltages:

$$\phi_0 e^{j\alpha_0} = C \sum_{p=1}^{7} \phi_p e^{j\alpha_p} \tag{2.128}$$

Consequently, this device constitutes a monolithic phasor adder. (The constant of proportionality C could be determined by use of Equation 2.126, but this is relatively unimportant.)

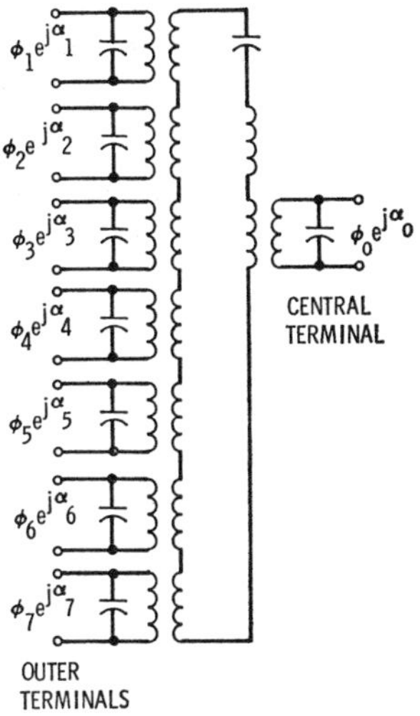

Figure 2.21. Equivalent circuit for the disk of the Figure 2.20 near one of the axisymmetric resonances.

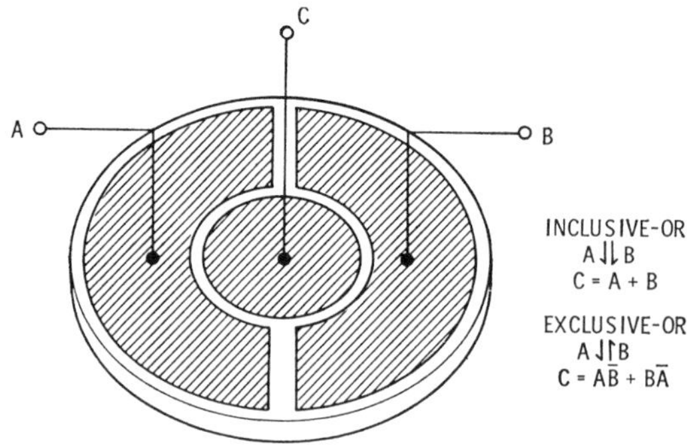

Figure 2.22. Ferroelectric ceramic disk logic element.

We have so far implicitly assumed that all the peripheral electroded regions in Figure 2.20 are equally polarized. However, if this is not the case, we will obtain a phasor adder with different weights assigned to the various inputs:

$$\phi_0 e^{j\alpha_0} = \sum_{p=1}^{7} C_p \phi_p e^{j\alpha_p} \qquad (2.129)$$

We should observe that these weight factors C_p are all continuously variable by direct electrical means; *i.e.*, by varying the polarization at electrode region p.

Let us now refer to the ferroelectric ceramic disk shown in Figure 2.22, where only two peripheral electrodes are present. In this case, if the two outer electrode regions are equally poled in the same direction, we obtain an inclusive-or gate: A signal at either A or B results in a signal at C. On the other hand, if the two outer electrode regions are poled equally but oppositely, we obtain an exclusive-or gate.[5]

Asymmetrically electroded ferroelectric ceramic disks may also be used to store information by associating memory states with degree or direction of polarization. For example, consider the eight-segment disk shown in Figure 2.23.

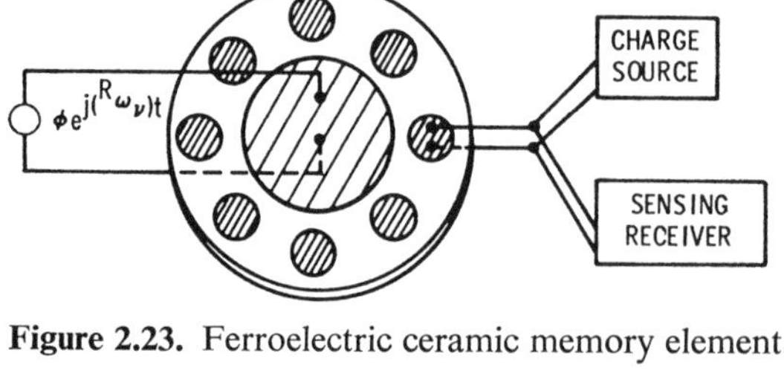

Figure 2.23. Ferroelectric ceramic memory element.

We shall first discuss the process of writing information into this device. Each of the eight outer electrode regions is equipped with a charge source for switching the polarization state in that region. (To avoid crowding, this charge source is only illustrated at one region.) It has been found experimentally[4] that an octal digit may be stored at each electrode pair. The eight states used to record this digit correspond to finite intervals rather than discrete values of residual polarization. For example, the first state may include values of polarization between 0 and $\frac{1}{8}$ maximum; the second, between $\frac{1}{8}$ and $\frac{2}{8}$ maximum, etc. In this manner, the disk shown in Figure 2.23 can store eight octal numbers.

The dependence of the piezoelectric coefficients on state of residual polarization is then used to obtain nondestructive, simultaneous readout of the eight polarization states. Let the central region be maximally polarized and supplied with a sinusoidal voltage of frequency corresponding to one of the angularly symmetric mode eigenfrequencies of the disk. This will keep the disk in continuous mechanical oscillation. Each outer region is thus piezoelectrically coupled to the mechanical vibrations of the disk by coefficients which depend on the polarization state at that region and hence on the octal digit stored there. Consequently, if each of the outer regions is provided with a receiver sensitive to radio frequency voltage amplitudes (sensing receiver), the signals picked up at each receiver may be readily related to the octal number stored at that region. (In Figure 2.23, we have illustrated the sensing receiver at only one of the eight regions in order to avoid crowding.)

External readout circuitry may be considerably simplified at the expense of information storage density if one only stores a binary rather than an octal digit at each electrode pair.

Because this memory device has a certain inherent dependence on the propagation velocity of sound across the peripheral electrode regions, it probably cannot compete in terms of cycle time with purely electronic storage elements. However, in applications where energy required for read-in, storage density, nonvolatility, and radiation insensitivity are more important than speed, a ferroelectric memory of this type is believed to be potentially quite feasible and useful.[4,5]

2.4. The Impedance of a Multilayered Plate

2.4.1. Device Applications and Other Observations

To conclude this chapter, we will now take up a somewhat different application of Green's functions: design and analysis of multilayer

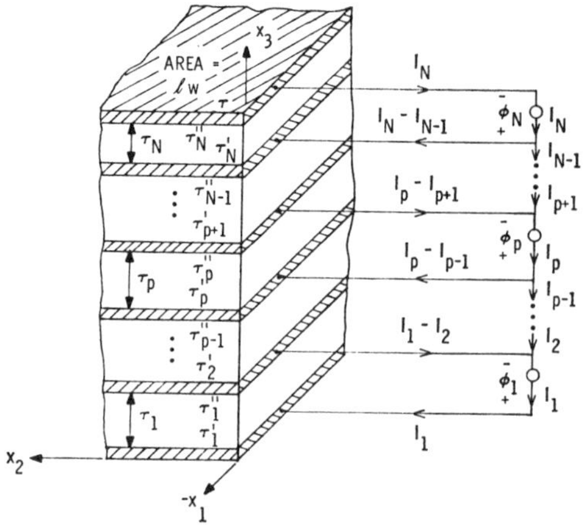

Figure 2.24. The N-layer problem. Cross-hatched regions represent electrode layers, and unmarked regions are ferroelectric ceramic layers. The sample is assumed to be free at $x_3 = 0$ and bonded to a substrate at $x_3 = \tau$.

piezoelectric plate devices. In particular, we will determine the impedance matrix of the arrangement shown in Figure 2.24.

One use for this type of structure was proposed by Newell in the mid 1960's.[10,62,63] He suggested mounting thickness mode piezoelectric resonators to a substrate by bonding them on acoustic quarter-wave plates of alternately high-Z and low-Z material. This scheme, in principle at least, achieves several desirable effects. First, proper arrangement of the quarter-wave plates makes the resonator appear to be floating nearly free in space — at least at the frequency of resonance. Consequently, the resonator is not damped significantly by its support. Second, the structure is mechanically strong and well supported with respect to random vibration or shock. Third, while the supporting quarter-wave plates do not suppress the desired thickness resonance, they will tend to suppress unwanted spurious responses such as high contour-extensional overtones of the plate (see Section 4.4). Consequently, this type of resonator usually has a cleaner response than one which actually *is* nearly floating unsupported in free space.

In practice, bonding difficulties and very critical tolerances plague this scheme of Newell's, and to date it has not proved effective in competing with "energy-trapped" thickness mode resonators discussed in Chapter 5. However, this multilayer plate configuration has found wide acceptance in multistage transducers, both for sonar[21,64,65] and delay line[66,67] applications.

2.4.2. Theory

Let us now refer back to the configuration shown in Figure 2.24, where alternate layers of electrode and piezoelectric material are present. The length l and width w of the sample will be assumed to be much greater than its thickness τ. In contrast to other problems treated in this chapter, excitation will now be at a frequency for which τ, rather than l or w, is on the order of an acoustic wavelength. Thus, if we assume the piezoelectric layers in Figure 2.24 are ferroelectric ceramic, u_3 will be the only significant displacement and S_3 the only strain. The inertia of the material acts as a mechanical clamp in the x_1x_2-plane, and constrains S_1, S_2, and S_6 to zero, while S_4 and S_5 are not piezoelectrically coupled to E_3.

For all other problems analyzed in this chapter there are more nonzero strains than stresses. Here, however, we have several nonzero stresses and just one nonzero strain. For problems in which fewer strains than stresses are present, the piezoelectric constituent relations are most conveniently expressed with S_μ rather than T_μ independent. In addition, in Figure 2.24 current rather than voltage is specified externally, so that D_m is preferable to E_m for the independent electrical variable. Thus, we now describe the linear macroscopic response of the ferroelectric ceramic layers by Equation 1.25,

$$
\begin{aligned}
T_3 &= c_{33}^D S_3 - h_{33} D_3 \\
E_3 &= -h_{33} S_3 + \beta_{33}^S D_3
\end{aligned}
\tag{2.130}
$$

The electrode layers of the sample have $E_3 = h_{33} = 0$, so that there the constituent relations become

$$
\begin{aligned}
T_3 &= c_{33}^D S_3 \\
0 &= D_3
\end{aligned}
\tag{2.131}
$$

In analogy with Equation 2.3, it is convenient to combine Equations 2.130 and 2.131 into a single set of equations, valid through the entire sample:

$$
\begin{aligned}
T_3 &= c_{33} S_3 - h_{33} D_3 \\
E_3 &= -h_{33} S_3 + \beta_{33}^S D_3
\end{aligned}
\tag{2.132}
$$

The following conventions are understood to apply in Equation 2.132:

$$D_3 = \begin{cases} I_{33}/(j\omega lw) & \text{in ceramic region } p \text{ (see Figure 2.24)} \\ 0 & \text{in all electrodes} \end{cases}$$

$$h_{33} = \begin{cases} h_{33}(x_3) & \text{in ceramic region } p \\ 0 & \text{in all electrodes} \end{cases}$$

$$\beta_{33}^S = \beta_{33}^S(x_3)$$

$$c_{33} = \text{a constant which is a weighted average of } c_{33}^D(x_3) \text{ and is defined analogously to } s_{11} \text{ in Equation 2.3} \tag{2.133}$$

The sample is assumed to be free at $x_3 = 0$ and bonded to the substrate at $x_3 = \tau$. Thus, the boundary conditions are

$$\begin{aligned} T_3\big|_{x_3=0} &= 0 \\ T_3\big|_{x_3=\tau} &= -j\omega\zeta_\tau u_3 \end{aligned} \tag{2.134}$$

where ζ_τ is the characteristic mechanical impedance of the substrate. We will assume the substrate is chosen so that its characteristic mechanical impedance is much lower than that of the ceramic. In other words, we will assume an acoustic mismatch is present at the ceramic-substrate interface which prevents much energy from escaping into the substrate.

Under these restrictions the particle displacement u_3 obeys the inhomogeneous wave equation,

$$\frac{\partial^2 u_3}{\partial x_3^2} + \frac{\rho\omega^2 u_3}{c_{33}} = \frac{1}{c_{33}} \frac{\partial (h_{33} D_3)}{\partial x_3} \tag{2.135}$$

and the boundary conditions

$$\begin{aligned} \frac{\partial u_3}{\partial x_3}\bigg|_{x_3=\epsilon} &= \frac{h_{33} D_3}{c_{33}}\bigg|_{x_3=\epsilon} \\ \frac{\partial u_3}{\partial x_3}\bigg|_{x_3=\tau-\epsilon} &= \left(\frac{h_{33} D_3}{c_{33}} - \frac{j\omega\zeta_\tau u_3}{c_{33}}\right)\bigg|_{x_3=\tau-\epsilon} \end{aligned} \tag{2.136}$$

The Green's function $G(z' \mid z'')$ which permits us to integrate this differential equation system must be defined to satisfy the differential equation

$$\frac{\partial^2 G(z' \mid z'')}{\partial z'^2} + \frac{\rho\omega^2}{c_{33}} G(z' \mid z'') = -\delta(z' - z'') \tag{2.137}$$

and the boundary conditions

$$\left.\frac{\partial G(z' \mid z'')}{\partial z'}\right|_{z'=\epsilon} = 0$$
$$\left.\frac{\partial G(z' \mid z'')}{\partial z'}\right|_{z'=\tau-\epsilon} = -\frac{j\omega\zeta_\tau}{c_{33}} \left. G(z' \mid z'')\right|_{z'=\tau-\epsilon} \tag{2.138}$$

There are many mathematical similarities between this problem and the N-segment bar problem analyzed in Section 2.1. In particular, Equations 2.132 to 2.138 are analogous to Equations 2.3 to 2.8. Thus, a manipulation identical to that carried out in Section 2.1 leads to the result

$$u_3(z'') = \int_{-\epsilon}^{\tau+\epsilon} \frac{1}{c_{33}} G(z' \mid z'') \frac{\partial(h_{33} D_3)}{\partial z'} dz' \tag{2.139}$$

which corresponds to Equation 2.14.

Equations 2.137 and 2.138 for $G(z' \mid z'')$ may be solved to give

$$G(z' \mid z'') = -\frac{2c_{33}}{\rho\tau} \sum_{\nu=1}^{\infty} \frac{(1 - \Delta_\nu/\nu) \cos[(\nu + \Delta_\nu)\pi z'/\tau] \cos[(\nu + \Delta_\nu)\pi z''/\tau]}{\omega^2 - \omega_\nu^2} \tag{2.140}$$

where

$$\Delta_\nu = \frac{j\zeta_\tau}{\pi\sqrt{c_{33}\rho}} \tag{2.141}$$

and

$$\omega_\nu = \frac{(\nu + \Delta_\nu)\pi}{\tau} \sqrt{\frac{c_{33}}{\rho}} \tag{2.142}$$

The quantity $\sqrt{c_{33}\rho}$ is the characteristic mechanical impedance of the ceramic, so that, in view of the postulated mismatch conditions Δ_ν is small compared to unity.

Now there can be no free charge inside ceramic slab p. Consequently, in that region the electrical connection shown in Figure 2.24 will require

$$D_3 = \text{constant} = \frac{I_p}{j\omega lw} \tag{2.143}$$

However, D_3 is zero in the electrode layers just outside the ceramic slab p. Hence, the driving term in the differential equation for u_3 may be evaluated as a series of force singularities at the ceramic-electrode interfaces to give

$$\frac{1}{c_{33}}\frac{\partial(D_3h_{33})}{\partial z'} = \frac{1}{j\omega lw}\sum_{p=1}^{N}\frac{h_{33}(p)I_p}{c_{33}}[\delta(z'-\tau_p') - \delta(z'-\tau_p'')] \quad (2.144)$$

Here τ_p' and τ_p'' are the coordinates of the x_3-boundary of ceramic region p, as Figure 2.24 indicates, and $h_{33}(p)$ is h_{33} evaluated in ceramic region p. This driving term is very similar to Equation 2.6, the corresponding driving term for the N-electrode bar. However, in this case the force singularities act at the electrode-ceramic interfaces rather than at the electrode edges and are proportional to $h_{33}(p)I_p$ rather than to $d_{31}(p)\phi_p$.

Substitution of Equations 2.140 and 2.144 into Equation 2.139 results in an explicit formula for the displacement function u_3:

$$u_3(z'') = \frac{1}{j\omega lw}\sum_{p=1}^{N}\sum_{\nu=1}^{\infty}\frac{h_{33}(p)I_p[u_3^{(\nu)}(\tau_p') - u_3^{(\nu)}(\tau_p'')]u_3^{(\nu)}(z'')}{\omega^2-\omega_\nu^2} \quad (2.145)$$

where

$$u_3^{(\nu)}(z'') = \sqrt{\frac{2}{\tau\rho}}\left(1-\frac{\Delta_\nu}{2\nu}\right)\cos\frac{(\nu+\Delta_\nu)\pi z''}{\tau} \quad (2.146)$$

The functions $u_3^{(\nu)}(z'')$ are actually the mechanical eigenfunctions of the homogeneous version of the elastic boundary value problem specified by Equations 2.135 and 2.136, and thus correspond to $u_1^{(\nu)}$ of Equation 2.61.

The voltage at terminals p may now be evaluated in terms of the driving currents and the properties of the plate:

$$\phi_p = \int_{\tau_p'}^{\tau_p''} E_3\,dx_3 = \int_{\tau_p'}^{\tau_p''}(-h_{33}(p)S_3 + \beta_{33}^S(p)D_3)\,dx_3 \quad (2.147)$$

where $\beta_{33}^S(p)$ is β_{33}^S evaluated in ceramic layer p.

As we previously remarked, there can be no free charge inside ceramic region q; inside that region Equation 2.143 may again be used:

$$D_3 = \text{constant} = \frac{I_q}{j\omega lw} \quad (2.148)$$

The S_3 component of the strain is defined as

$$\frac{\partial u_3}{\partial x_3} = S_3 \quad (2.149)$$

Substitution of these two relations into Equation 2.147 then yields

$$\phi_p = \frac{I_p}{j\omega C_0(p)} - h_{33}(p)[u_3(\tau_p'') - u_3(\tau_p')] \tag{2.150}$$

where

$$C_0(p) = \frac{lw}{\tau_p \beta_{33}^S(p)} \tag{2.151}$$

is the clamped capacitance of ceramic layer p.

If, finally, Equation 2.145 for $u_3(z'')$ is substituted into Equation 2.150, and if the result is differentiated with respect to I_q, we get the desired impedance matrix of the device shown in Figure 2.24,

$$Z_{pq} = \frac{\partial \phi_p}{\partial I_q} = \frac{\delta_{pq}}{j\omega C_0(p)} + \frac{h_{33}(p)h_{33}(q)}{j\omega lw\rho} \cdot \frac{2}{\tau} \sum_{\nu=1}^{\infty} \left(1 - \frac{\Delta_\nu}{\nu}\right) \cdot \frac{1}{\omega^2 - \omega_\nu^2}$$
$$\cdot \left[\cos \frac{\tau_p'(\nu + \Delta_\nu)\pi}{\tau} - \cos \frac{\tau_p''(\nu + \Delta_\nu)\pi}{\tau}\right]$$
$$\cdot \left[\cos \frac{\tau_q'(\nu + \Delta_\nu)\pi}{\tau} - \cos \frac{\tau_q''(\nu + \Delta_\nu)\pi}{\tau}\right] \tag{2.152}$$

This impedance matrix is extremely similar in form to the admittance matrix found in Section 2.1 for the N-segment bar. In many respects the N-layer plate and the N-segment bar are dual problems.

It may often occur in the case of multilayer plate problems that the mismatch between electrode and piezoelectric layers is too great to make valid the uniform c_{33} approximation of Equations 2.132 and 2.133. (This will be the case for mismatches greater than about 25 percent.) It may also happen that we wish to treat multilayer problems with independent mechanical drives applied at $x_3 = 0, \tau$, or problems loaded so heavily that the ratio of Equation 2.141 is not much smaller than unity. These situations are especially likely to occur in designing transducers. The concepts presented in this section have elsewhere been generalized to include those cases.[68,69] Equivalent circuit representations of Z_{pq} as given by Equation 2.152 have also been published elsewhere.[68] The equivalent circuit for Z_{pq} is, unfortunately, considerably more complicated than those shown earlier in the chapter for the various Y_{pq}'s, and consequently will not be included here.

3 Variational Elasticity Theory

There are two primary shortcomings of the Green's function technique presented in the previous chapter. First, it is not applicable to piezoelectric problems that are essentially three- rather than one- or two-dimensional. Second, it presupposes a knowledge of the normal or free modes of vibration of the sample.

In practice, relatively few piezoelectric devices are basically three-dimensional, and most devices are constructed in a geometry for which the normal modes may simply be found by solving ordinary differential equations. Consequently, the two difficulties mentioned here do not present as much of a problem as might, at first, be anticipated.

However, there are some piezoelectric devices in which basically three-dimensional motions do take place. Moreover, there are two-dimensional cases in which one cannot, by any known technique, determine the normal modes exactly. These situations require the use of procedures other than those presented in Chapter 2. This and the next chapter will be devoted to the development of variational techniques, some exact and some approximate, which are universally applicable to any problem of linear electroelasticity. By means of these variational techniques, we can deal successfully with those device classes that defy Green's function analysis.

In the present chapter, we shall review the basic concepts of variational calculus as used to approximate the normal modes of ordinary (nonpiezoelectric), linearly elastic structures. These concepts have been known for a long time and are often referred to as "Rayleigh procedures" or "Rayleigh-Ritz techniques."[70,71,72] In the next chapter, we shall present a recently developed method of extending these variational techniques to evaluate the normal modes of three-dimensional samples with piezoelectric effects present. We can additionally use the calculus of variations to derive formulas relating admittance and impedance matrices of three-dimensional piezoelectric structures to their normal mode characteristics. These three-dimensional formulas, which also are described in the next chapter, may be regarded as generalizations of the two-dimensional results obtained in Chapter 2.

3.1. Lagrangian Formulation for Determination of Elastic Normal Modes

Let us begin by considering an arbitrarily shaped, mechanically free, three-dimensional elastic structure occupying a volume V. It will be the purpose of this section to describe a variational procedure for approximating the acoustic eigenmodes of this structure.

The electroelastic constituent relations, Equations 1.28, for a material of arbitrary crystallographic symmetry in the nonpiezoelectric case reduce to

$$T_{ij} = c_{ijkl}S_{kl} \tag{3.1}$$

and a similar dielectric equation which is not relevant here. We shall assume linear but not necessarily homogeneous conditions and an $e^{j\omega t}$ time dependence to prevail. If particle displacement is represented as before by u_k, Equation 3.1 may be rewritten

$$T_{ij} = c_{ijkl}u_{k,l} \tag{3.2}$$

where commas, as before, denote covariant differentiation when Equation 3.2 is applied in non-Cartesian coordinates.

The momentum conservation equation for a differential volume element now takes the form

$$T_{ij,j} = (c_{ijkl}u_{k,l})_{,j} = -\rho\omega^2 u_i \tag{3.3}$$

Let us specify the surface of the sample by ∂V. The postulated mechanically free boundary condition on ∂V is then expressed mathematically as

$$N_j T_{ij} = N_j c_{ijkl}u_{k,l} = 0 \tag{3.4}$$

The problem of formulating this situation variationally is now a matter of finding some Lagrangian expression which is stationary, *i.e.*, has a zero first variation, *only if* Equations 3.3 and 3.4 hold.[73] The following expression may be shown to meet this requirement:

$$L = \iiint_V \tfrac{1}{2}(u_{i,j}c_{ijkl}u_{k,l} - \rho\omega^2 u_i u_i)\, dV \tag{3.5}$$

We shall now demonstrate the above claim regarding this expression. To compute the first variation δL, assign arbitrary independent variations to u_i and its derivative $u_{i,j}$. Then evaluate the corresponding first-order change in L. This change is the desired δL,

$$\delta L = \iiint_V (\delta u_{i,j}c_{ijkl}u_{k,l} - \rho\omega^2\, \delta u_i u_i)\, dV \tag{3.6}$$

The variation in u_i may be any infinitesimal function possessing second derivatives; it need not obey the boundary condition Equation 3.4 or differential equation 3.3.

If Equation 3.6 for δL is rearranged by use of the divergence theorem, we may obtain

$$\delta L = -\iiint_V \delta u_i[(c_{ijkl}u_{k,l})_{,j} + \rho\omega^2 u_i]\, dV + \oiint_{\partial V} \delta u_i[N_j c_{ijkl}u_{k,l}]\, dA \tag{3.7}$$

As the δu_i was specified to be arbitrary, this δL can vanish irrespective of δu_i *only if* the two terms in brackets are zero everywhere in their domains of integration. However, this requirement corresponds exactly to the problem formulated by Equations 3.3 and 3.4. Consequently, the Lagrangian of Equation 3.5 is appropriate for the problem at hand.

We utilize this Lagrangian by approximating the u_i with a linear combination of trial functions $U_i^{(\alpha)}$,

$$u_i \doteq \sum_{\alpha=1}^{S_m} B^{(\alpha)} U_i^{(\alpha)} \tag{3.8}$$

and substituting that approximation in Equation 3.5. Here the $B^{(\alpha)}$ are undetermined coefficients to be evaluated. The trial functions $U_i^{(\alpha)}$ should all possess second derivatives. However, they need not satisfy the differential equation or boundary condition, except that they should not be zero for all α at any point in V or on ∂V where the exact solutions are not anticipated to be zero.*

If the $U_i^{(\alpha)}$ form a complete set of functions that do not violate the earlier-mentioned nonzero restriction, the method we are in the process of describing will give solutions which are convergent to the exact answers as S_m goes to infinity.†

Upon substituting Equation 3.8 into 3.5, we get the following approximation for L:

$$L \doteq \frac{1}{2} \sum_{\alpha=1}^{S_m} \sum_{\beta=1}^{S_m} B^{(\alpha)} B^{(\beta)} [P(\alpha,\beta) - \omega^2 H(\alpha,\beta)] \tag{3.9}$$

In Equation 3.9, the P is a matrix representing elastic energy interaction between trial functions,

$$P(\alpha,\beta) = \iiint_V U_{i,j}^{(\alpha)} c_{ijkl} U_{k,l}^{(\beta)} \, dV \tag{3.10}$$

and the H is a matrix representing kinetic energy interaction,

$$H(\alpha,\beta) = \iiint_V U_i^{(\alpha)} \rho U_i^{(\beta)} \, dV \tag{3.11}$$

The $B^{(\alpha)}$ coefficients may now be determined from the condition that the Lagrangian be stationary,

$$\frac{\partial L}{\partial B^{(\gamma)}} = \sum_{\alpha=1}^{S_m} B^{(\alpha)} [P(\alpha,\gamma) - \omega^2 H(\alpha,\gamma)] = 0 \qquad \gamma = 1, 2, \ldots, S_m \tag{3.12}$$

This operation is analogous to letting $\delta u_i = U_i^{(\gamma)} \delta B^{(\gamma)}$ in Equation 3.7. Consequently, the statement which was made after Equation 3.6 that δu_i need not obey the boundary conditions or differential equations gives us the liberty mentioned above to let $U_i^{(\gamma)}$ violate the boundary conditions and differential equations. As with δu_i, the $U_i^{(\gamma)}$ do need to have second derivatives in V and on ∂V, though.

* If, however, the $U_i^{(\alpha)}$ are selected so that the boundary condition or differential equation is satisfied, one will obtain an enhanced rate of convergence.

† Sets of functions that are complete in the ordinary Fourier or normwise sense and are not all zero at the same point are said to be pointwise complete.[74] Thus, $\{\cos mx\}$ forms a pointwise complete set on $0 \leq x \leq \pi$, but $\{\sin mx\}$ is only normwise complete. In most cases, our work will require trial function sets that are pointwise complete.

Equation 3.12 is very amenable to expression in matrix form,

$$[P - \omega^2 H][B] = [0] \tag{3.13}$$

As P and H are symmetric matrices, Equation 3.13 constitutes a symmetric characteristic value problem. It will only possess nontrivial solutions for discrete values of ω^2. These eigenvalues, of course, give the approximate resonant frequencies squared ω_ν^2 for the elastic structure in question.

We shall find it convenient to describe the eigenvectors associated with the ω_ν by $[B^{(\nu)}]$, and the individual components of $[B^{(\nu)}]$ by $B^{(\alpha,\nu)}$. These eigenvectors are determined only to a multiplicative constant by Equation 3.13. The corresponding approximate modal displacement patterns are specified by substituting the $B^{(\alpha,\nu)}$ into Equation 3.8,

$$u_i^{(\nu)} \doteq \sum_{\alpha=1}^{S_m} B^{(\alpha,\nu)} U_i^{(\alpha)} \tag{3.14}$$

It is known that the exact $u_i^{(\nu)}$ associated with different ω_ν are orthogonal with respect to ρ in both the two-dimensional case, Equation 2.63, and the present three-dimensional case.[55] This is also true of the present approximate solutions given by Equation 3.14.[32] For subsequently described applications, or for use in connection with the theory presented in Section 2.3, it is highly desirable, however, to have these approximate displacement patterns orthonormal rather than merely orthogonal. This may be most easily effected by generalizing the normalization condition, Equation 2.63, to three dimensions,

$$\iiint_V u_i^{(\nu)} \rho u_i^{(\mu)}\, dV = \delta_{\mu\nu} \tag{3.15}$$

If Equation 3.14 is substituted into Equation 3.15, we obtain the normalization condition on the $[B^{(\nu)}]$,

$$[B_t^{(\nu)}][H][B^{(\mu)}] = \delta_{\mu\nu} \tag{3.16}$$

Here the definition of $[H]$ from Equation 3.11 has been applied, and $[B_t^{(\nu)}]$ is the transpose of the vector $[B^{(\nu)}]$.

3.2. Applications to Thick Disks

In this section, we shall use the variational techniques just described to determine the angularly symmetric dilational modes and resonant

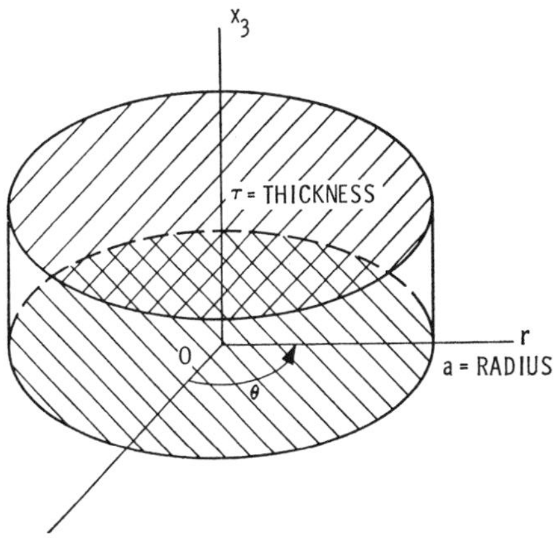

Figure 3.1. An elastic disk of comparable thickness and diameter $2a$, with coordinate system.

frequencies of the elastic disk shown in Figure 3.1. Comparable thickness and diameter dimensions will be assumed here, so that the negligible thickness restrictions of Section 2.2 are not appropriate. The disk will be treated as isotropic in its own plane but anisotropic along its axis, so that it will behave as a ferroelectric ceramic without piezoelectric effects. In other words, the independent elastic coefficients will be

$$c_{1111} = c_{2222},\ c_{1122},\ c_{1133} = c_{2233},\ c_{3333},\ c_{1212},\ c_{1313} = c_{2323} \tag{3.17}$$

This problem has no apparent practical application, except that it will serve as an interesting contrast in the next chapter when we treat the same thick-disk geometry with piezoelectric effects *present*. (There are a number of applications for this second case in which one has piezoelectric coupling.)

As we are going to consider only the angularly symmetric dilation modes, we may use trial functions with no θ-component, $U_\theta^{(\alpha)} = 0$, and no θ-dependence of $U_r^{(\alpha)}$ and $U_3^{(\alpha)}$. Moreover, we may assume the $U_r^{(\alpha)}$ are all even in x_3 and the $U_3^{(\alpha)}$ are all odd in x_3. In other words, it is permissible here and in all other cases to restrict the trial functions so they have the same symmetry as the desired solution.

In particular, the following two-dimensional series of trial functions satisfies the symmetry requirements for this geometry and is normwise complete:

$$\left.\begin{aligned} U_r^{(\alpha)} &= J_1(h_\alpha r/a)\sin[(2m_\alpha - 1)\pi x_3/\tau] \\ U_3^{(\alpha)} &= 0 \end{aligned}\right\} \qquad \alpha = 1, 2, \ldots, S_1$$

$$\left.\begin{aligned} U_r^{(\alpha)} &= 0 \\ U_3^{(\alpha)} &= J_0(k_\alpha r/a)\cos[(2n_\alpha - 1)\pi x_3/\tau] \end{aligned}\right\} \qquad \alpha = S_1 + 1, S_1 + 2, \ldots, S_m \tag{3.18}$$

Here m_α and n_α run 1,2, . . . and h_α are the roots of

$$h_\alpha J_0(h_\alpha) = (1 - c_{1122}/c_{1111})J_1(h_\alpha) \tag{3.19}$$

and k_α are the roots of

$$J_0(k_\alpha) = 0 \tag{3.20}$$

In these equations, the m_α, n_α, h_α, and k_α are assigned subscripts running 1 to S_m so that these two-dimensional expansions can be transformed into the one-dimensional series required by Equation 3.8.

These trial functions are selected primarily because they prove to be easy to manipulate in computing the $H(\alpha,\beta)$ and $P(\alpha,\beta)$ matrices of Equations 3.10 and 3.11. They do not, individually, bear any particularly close resemblance to the exact eigenmodes.

Basically, there are two ways to approach the task of selecting trial functions for variational procedures:

(1) We can choose trial functions that do resemble the exact solutions — for instance, they may be tailored to satisfy either the differential equation or boundary condition. This approach has the advantage of giving rapid convergence with relatively few trial functions. However, tailored trial functions are usually quite complicated and make the evaluation of $P(\alpha,\beta)$ and $H(\alpha,\beta)$ very tedious.

(2) We can, as we do here, use trial functions that are not in any way refined but are simple to work with. This approach will yield slower convergence, but in the era of high-speed computers, no grave problems are generated by that. In effect, letting the trial functions be unrefined places more of the computational burden on the machine and less on the mathematician.

However, when unsophisticated sets of trial functions are used, it is vital that the pointwise-completeness requirement described in the previous section be obeyed. The x_3-dependences of our present choice of functions are sine-cosines and thus clearly normwise complete. The r-dependences also are normwise complete, as they represent the exact eigenmodes of certain thin elastic disk or circular membrane Sturm-Liouville systems.[43] Such eigenmodes are known to be normwise complete.[54]

However, our trial functions in Equations 3.18 are not pointwise complete: All the $U_r^{(\alpha)}$ are zero at $x_3 = 0, \tau$, and all the $U_3^{(\alpha)}$ are zero at $r = a$. Consequently, if substituted into Equations 3.8 to 3.16 of the previous section, these functions do not yield satisfactory results. Moreover, it is quite difficult to find alternative simple sets of trial functions which do have the property of pointwise completeness and yet retain the computational advantages of Equation 3.18.

There is, fortunately, a good compromise way out of this dilemma. In particular, we may add additional corrective terms to the expansion of Equation 3.18 which make the resultant expansion "over complete" in a normwise sense but also give it the desired pointwise completeness. Deciding what types of over-complete terms to add may occasionally require some trial and error or insight. Usually, though, almost any reasonable-looking addition which has elements nonzero at those points where all the elements of the main expansion, Equations 3.18, are zero will work.

In this case, we have personally had excellent success by using the over-complete modification of Equation 3.18,

$$\left.\begin{aligned} U_r^{(\alpha)} &= J_1(h_\alpha r/a)\sin[(2m_\alpha - 1)\pi x_3/\tau] \\ U_3^{(\alpha)} &= 0 \end{aligned}\right\} \qquad \alpha = 1, 2, \ldots, S_1$$

$$\left.\begin{aligned} U_r^{(\alpha)} &= J_1(h_\alpha r/a) \\ U_3^{(\alpha)} &= 0 \end{aligned}\right\} \qquad \alpha = S_1 + 1, \quad S_1 + 2, \ldots, S_2$$

$$\left.\begin{aligned} U_r^{(\alpha)} &= 0 \\ U_3^{(\alpha)} &= J_0(k_\alpha r/a)\cos[(2n_\alpha - 1)\pi x_3/\tau] \end{aligned}\right\} \qquad \alpha = S_2 + 1, \quad S_2 + 2, \ldots, S_3$$

$$\left.\begin{aligned} U_r^{(\alpha)} &= 0 \\ U_3^{(\alpha)} &= \cos[(2n_\alpha - 1)\pi x_3/\tau] \end{aligned}\right\} \qquad \alpha = S_3 + 1, \quad S_3 + 2, \ldots, S_m \tag{3.21}$$

Again m_α and n_α run 1, 2, . . . and h_α and k_α are the roots of Equations 3.19 and 3.20 respectively. It should be noted that the over-completional corrective series $U_r^{(\alpha)} = J_1(h_\alpha r/a)$ and $U_3^{(\alpha)} = \cos[(2n_\alpha - 1)\pi x_3/\tau]$ are merely one-dimensional in character. For most instances it is sufficient, and in fact even desirable, to make the corrective series one-dimensional.

In the next section, we will once more encounter the necessity for over-complete expansions. At that time, we will present an explanation for the desirability of making the corrective series one-dimensional.

In cylindrical coordinates with $U_\theta^{(\alpha)} = \partial/\partial\theta = 0$, the general forms for the interaction integral matrices $P(\alpha,\beta)$ and $H(\alpha,\beta)$ of Equations 3.10 and 3.11 are

$$\begin{aligned} P(\alpha,\beta) = \iiint_V \Bigg[& \frac{\partial U_r^{(\alpha)}}{\partial r}\left(c_{1111}\frac{\partial U_r^{(\beta)}}{\partial r} + c_{1122}\frac{U_r^{(\beta)}}{r}\right) \\ & + \frac{U_r^{(\alpha)}}{r}\left(c_{1122}\frac{\partial U_r^{(\beta)}}{\partial r} + c_{1111}\frac{U_r^{(\beta)}}{r}\right) + \frac{\partial U_r^{(\alpha)}}{\partial x_3}c_{1313}\frac{\partial U_r^{(\beta)}}{\partial x_3} \\ & + \frac{\partial U_3^{(\alpha)}}{\partial x_3}c_{1133}\left(\frac{\partial U_r^{(\beta)}}{\partial r} + \frac{U_r^{(\beta)}}{r}\right) + \frac{\partial U_3^{(\alpha)}}{\partial r}c_{1313}\frac{\partial U_r^{(\beta)}}{\partial x_3} \\ & + \frac{\partial U_3^{(\beta)}}{\partial x_3}c_{1133}\left(\frac{\partial U_r^{(\alpha)}}{\partial r} + \frac{U_r^{(\alpha)}}{r}\right) + \frac{\partial U_3^{(\beta)}}{\partial r}c_{1313}\frac{\partial U_r^{(\alpha)}}{\partial x_3} \\ & + \frac{\partial U_3^{(\alpha)}}{\partial x_3}c_{3333}\frac{\partial U_3^{(\beta)}}{\partial x_3} + \frac{\partial U_3^{(\alpha)}}{\partial r}c_{1313}\frac{\partial U_3^{(\beta)}}{\partial r}\Bigg]\, dV \end{aligned} \tag{3.22}$$

and

$$H(\alpha,\beta) = \iiint_V [U_r^{(\alpha)}\rho U_r^{(\beta)} + U_3^{(\alpha)}\rho U_3^{(\beta)}]\, dV \tag{3.23}$$

In order to derive the formula for $P(\alpha,\beta)$, the covariant derivative $U_{k,l}^{(\alpha)}$ must be either computed in polar coordinates with the use of Christoffel symbols[75] or looked up in a table.[76,77] The elastic symmetry conditions, Equations 3.17, have also been incorporated into Equation 3.22.

Recursion formulas for the elements of these matrices as obtained from the trial function of Equation 3.21 have been published elsewhere,[78] along with a computer program for evaluating those recursion formulas and solving the resulting characteristic value equation, Equation 3.12.

In the course of this research, actual numerical values have been computed for thick disks characterized by the c_{ijkl} and ρ values given

Table 3.1. Elastic c^E_{ijkl} Coefficients and ρ of Clevite Ceramic A ($BaTiO_3$) in MKS Units[17]

$c_{1111} = 1.50 \times 10^{11}$	$c_{1212} = 0.43 \times 10^{11}$
$c_{1122} = 0.66 \times 10^{11}$	$c_{1313} = 0.44 \times 10^{11}$
$c_{1133} = 0.66 \times 10^{11}$	$\rho = 5.7 \times 10^{3}$
$c_{3333} = 1.46 \times 10^{11}$	

in Table 3.1. These c_{ijkl} and ρ values are chosen so as to coincide with c^E_{ijkl} and ρ of Clevite Ceramic A* ($BaTiO_3$), which is ferroelectric. Thus we may regard these results as describing the behavior of Ceramic A disks when piezoelectric effects are neglected. In Chapter 4, disks of the same material will be analyzed again, but then we will include piezoelectric effects, so that one can see how those effects modify the solution. Our reason for selecting Ceramic A for these computations is that Shaw has performed extensive experimental measurements on disks of that material.[79] This provides an opportunity to make an interesting check on our work, both when piezoelectricity is included and when it is not.

In performing these computations, a total of 30 trial functions were used: In the notation of Equation 3.21, we took

$$\begin{aligned} S_1 &= 12 \text{ with } (h_\alpha,m_\alpha) = (h_1,1), (h_2,1), (h_3,1), (h_4,1), (h_1,2), \\ &\qquad (h_2,2), (h_3,2), (h_4,2), (h_1,3), (h_2,3), \\ &\qquad (h_3,3), (h_4,3) \\ S_2 - S_1 &= 6 \text{ with } h_\alpha = h_1, h_2, h_3, h_4, h_5, h_6 \\ S_3 - S_2 &= 8 \text{ with } (k_\alpha,n_\alpha) = (k_1,1), (k_2,1), (k_3,1), (k_4,1), \\ &\qquad (k_1,2), (k_2,2), (k_3,2), (k_4,2) \\ S_m - S_3 &= 4 \text{ with } n_\alpha = 1, 2, 3, 4 \end{aligned} \tag{3.24}$$

Upon substituting these values in Equations 3.21, inserting the result in Equations 3.22 and 3.23, and then solving Equation 3.12 we get the spectrum shown in Figure 3.2. Also illustrated there are Shaw's experimental values. While the agreement between theory and experiment is clearly excellent, it may be observed that the theoretical resonant frequencies consistently fall slightly below the experimental frequencies. In the next chapter, we shall see that this effect is because of our omission here of piezoelectric corrections.

Several additional observations regarding Figure 3.2 are in order at this time. First, the resonant frequencies were multiplied by disk

* Clevite Corporation, trade name.

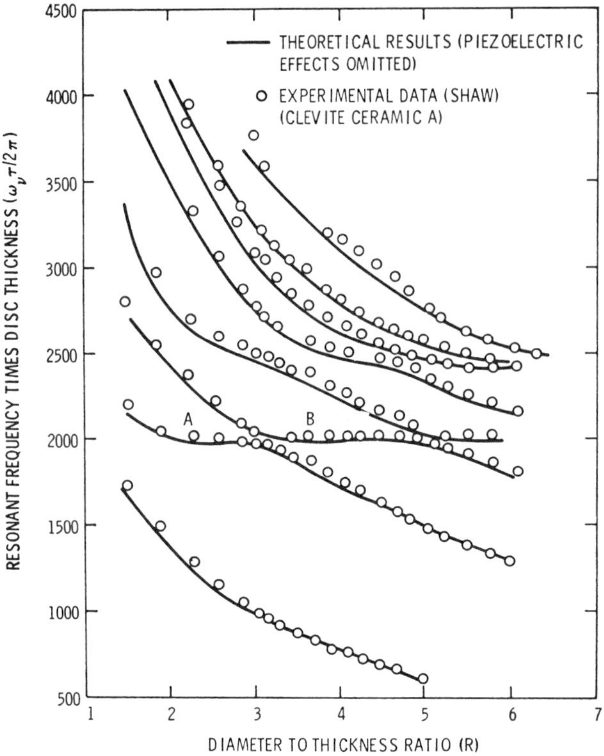

Figure 3.2. Comparison of theoretical and experimental resonant frequencies of thick ferroelectric ceramic disks. Piezoelectric effects are omitted in the theory here. Note that the theoretical values consistently are slightly low.

thickness τ before being plotted. Also the horizontal axis represents the only remaining free parameter in the problem, diameter-to-thickness ratio,

$$R = 2a/\tau \tag{3.25}$$

Consequently, the spectrum shown in Figure 3.2 is representative of any disk characterized by the coefficients of Table 3.1, irrespective of its absolute dimensions.

The portion of the spectrum labeled A and B in Figure 3.2 corresponds to the so-called edge mode in which mechanical displacements are primarily localized at the circumference of the disk. In the past, this mode has been described by the use of incomplete sets of trial functions with complex wavenumbers k_α and h_α.[80] However, because of the pointwise-complete nature of our trial functions, Equation 3.21, complex wavenumbers are not required here. (The introduction of complex wavenumbers, of course, would greatly enhance the edge-mode computational accuracy obtainable from a fixed number of trial functions: These complex-wavenumber trial functions are the result of a considerable amount of pretailoring to the needs of that particular problem.)

3.3. Applications to Rectangular Plates

3.3.1. Reduction of the General Theory to Two Dimensions

In order to apply the thin plate Green's function techniques presented in the previous chapter, Section 2.3, we must have a knowledge of the eigenmode solutions of Equation 2.61,

$$\begin{aligned} (\gamma^E_{ijkl} u^{(v)}_{k,l})_{,i} + \rho\omega_v^2 u_j^{(v)} &= 0 \qquad \text{over plate } A \\ N_j \gamma^E_{ijkl} u^{(v)}_{k,l} &= 0 \qquad \text{at plate edge } \partial A \\ i, j, k, l &= 1 \text{ and } 2 \text{ only} \end{aligned} \tag{3.26}$$

These equations, of course, govern the behavior of a mechanically free, thin piezoelectric plate completely covered by short-circuited electrodes (see Figure 2.10). Alternatively, if the superscript E is dropped from the γ's, they govern a thin nonpiezoelectric free plate.

There are a number of instances in which Equations 3.26 apparently cannot be solved exactly. The ferroelectric ceramic rectangular plate shown in Figure 3.3 is a case in point. Consequently, in dealing with

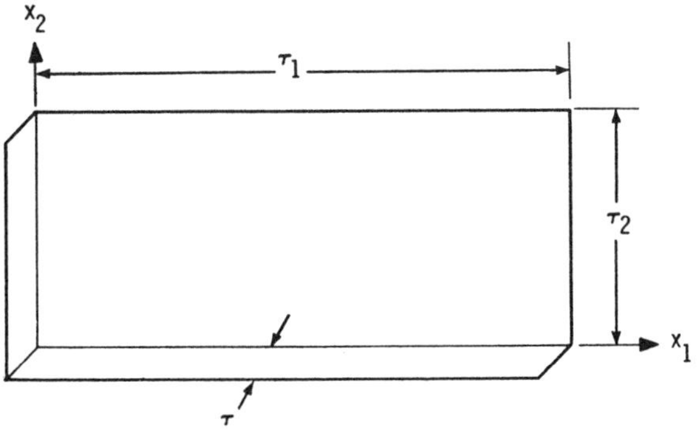

Figure 3.3. A thin rectangular ferroelectric ceramic plate with dimensions and coordinates. Polarization is in the τ direction.

this type of problem, it is necessary to resort to approximation techniques such as those presented in the first part of this chapter.

Equations 3.26 are identical to Equations 3.3 and 3.4, except that now c is replaced by γ^E and all subscripts run over 1 and 2 only. Consequently, in analogy with Equation 3.5, Equations 3.26 may be represented by the stationary Lagrangian

$$L = \iint_A \tfrac{1}{2}(u_{i,j}\gamma^E_{ijkl}u_{k,l} - \rho\omega^2 u_i u_i)\, dA \tag{3.27}$$

Here A represents the portion of the x_1x_2-plane covered by the plate.

This Lagrangian is utilized in the same manner as its three-dimensional sibling: Expand u_i in a linear combination of trial functions

$$u_i \doteq \sum_{\alpha=1}^{S_m} B^{(\alpha)} U_i^{(\alpha)} \tag{3.28}$$

and substitute in Equation 3.27. It is again required that the trial function set be pointwise complete and twice differentiable over the domain of interest, although the $U_i^{(\alpha)}$ need not obey either of Equations 3.26. The result of carrying out substitution of Equation 3.28 into Equation 3.27 is

$$L \doteq \frac{1}{2} \sum_{\alpha=1}^{S_m} \sum_{\beta=1}^{S_m} B^{(\alpha)}B^{(\beta)}[P(\alpha,\beta) - \omega^2 H(\alpha,\beta)] \tag{3.29}$$

where $P(\alpha,\beta)$ is the elastic energy interaction matrix

$$P(\alpha,\beta) = \iint_A U_{i,j}^{(\alpha)}\gamma^E_{ijkl}U_{k,l}^{(\beta)}\, dA \tag{3.30}$$

and $H(\alpha,\beta)$ is the kinetic energy interaction matrix,

$$H(\alpha,\beta) = \iint_A U_i^{(\alpha)}\rho U_j^{(\beta)}\, dA \tag{3.31}$$

We again evaluate the $B^{(\alpha)}$ coefficients by appealing to the stationary property of the Lagrangian,

$$\frac{\partial L}{\partial B^{(\gamma)}} = \sum_{\alpha=1}^{S_m} B^{(\alpha)}[P(\alpha,\gamma) - \omega^2 H(\alpha,\gamma)] = 0 \qquad \gamma = 1, 2, \ldots, S_m \tag{3.32}$$

The eigenvalues of Equation 3.32 give the resonant frequencies of the plate, and the associated eigenvectors represent the corresponding displacement patterns,

$$u_i^{(\nu)} = \sum_{\alpha=1}^{S_m} B^{(\alpha,\nu)} U_i^{(\alpha)} \tag{3.33}$$

These modal displacement patterns may be normalized in accordance with Equation 2.63 by normalizing the $B^{(\alpha,\nu)}$ eigenvectors as

$$[B_t^{(\nu)}][H][B^{(\mu)}] = \delta_{\mu\nu}/\tau \tag{3.34}$$

In Equation 3.34, the $[B^{(\nu)}]$ represents the array of S_m coefficients $B^{(\alpha,\nu)}$. The definition of $[H]$, Equation 3.31, is used in deriving Equation 3.34 from Equation 2.63.

It should, perhaps, be emphasized again that not only are the exact eigenfunctions of Equations 3.26 orthogonal to each other, but the approximate eigenfunctions as computed from Equation 3.33 are orthogonal also.[32]

3.3.2. *Considerations Governing the Choice of Trial Functions*

Let us now proceed to evaluate the resonant frequencies and eigenmodes of the rectangular ferroelectric ceramic plate shown in Figure 3.3. The relevant independent elastic coefficients are

$$\gamma_{1111}^E = \gamma_{2222}^E,\ \gamma_{1122}^E,\ \gamma_{1212}^E = \tfrac{1}{2}(\gamma_{1111}^E - \gamma_{1122}^E) \tag{3.35}$$

We shall restrict our attention to the dilation-type contour extensional modes, which are characterized by

$$\begin{array}{ll} u_1^{(\nu)} & \text{odd in } x_1 \text{ and even in } x_2 \text{ with respect to the center of } A \\ u_2^{(\nu)} & \text{even in } x_1 \text{ and odd in } x_2 \text{ with respect to the center of } A \end{array} \tag{3.36}$$

Other symmetry types of rectangular plate modes, including diagonal-shear and flexure, have been described elsewhere.[72,81,82,83,84]

Trial functions, which form a pointwise complete set, are easy to work with and have the symmetry of Equation 3.36 are given by

$$\left.\begin{aligned} U_1^{(\alpha)} &= \cos(m_\alpha \pi x_1/\tau_1)\cos(n_{c\alpha}\pi x_2/\tau_2) \\ U_2^{(\alpha)} &= 0 \end{aligned}\right\} \quad \alpha = 1, 2, \ldots, S_1$$

$$\left.\begin{aligned} U_1^{(\alpha)} &= 0 \\ U_2^{(\alpha)} &= \cos(m'_{c\alpha}\pi x_1/\tau_1)\cos(n'_\alpha \pi x_2/\tau_2) \end{aligned}\right\} \quad \alpha = S_2 + 1,\ S_2 + 2, \ldots, S_3 \tag{3.37}$$

Here, m_α and n'_α are odd integers 1, 3, 5, . . . while $m'_{c\alpha}$ and $n_{c\alpha}$ are even 0, 2, 4, ...

These trial functions, when substituted into Equations 3.30 to 3.34, do yield solutions convergent to the exact modes and resonant frequencies as more and more trial functions are added. However, this convergence is rather slow and even taking 100 terms may not yield 5 percent accuracy for the lowest ω_ν.

The reason for this behavior apparently lies in the fact that it takes a very large number of these functions to approximate the boundary condition of Equation 3.26 at all closely. These functions have a zero gradient on the entire boundary contour. While this zero gradient is a fair approximation to the actual requirement on u_1 at $x_1 = 0, \tau_1$ and on u_2 at $x_2 = 0, \tau_2$, it is a very poor approximation for u_1 at $x_2 = 0, \tau_2$ and for u_2 at $x_1 = 0, \tau_1$. In fact, the latter requirement is not well satisfied by trial functions which either are zero themselves or have a zero gradient on the boundary.

One satisfactory way of approaching this difficulty may be borrowed from the previous section. That approach is to add a corrective series of trial functions to Equations 3.37, so that the overall expansion becomes over complete. We stated previously that the apparent difficulty with the trial functions of Equations 3.37 lies in their inability to represent the nonzero gradient boundary condition governing u_1 at $x_2 = 0, \tau_2$ and u_2 at $x_1 = 0, \tau_1$. Consequently, the corrective series must be selected with the intent of remedying this difficulty. A corrective series which gives u_1 a nonzero gradient at $x_2 = 0, \tau_2$ and u_2 a nonzero gradient at $x_1 = 0, \tau_1$, but still possesses the symmetry dictated by Equation 3.36 is

$$\left.\begin{aligned} U_1^{(\alpha)} &= \cos(m_\alpha \pi x_1/\tau_1)\sin(n_{s\alpha}\pi x_2/\tau_2) \\ U_2^{(\alpha)} &= 0 \end{aligned}\right\} \qquad \alpha = S_1 + 1, \quad S_1 + 2, \ldots, S_2$$

$$\left.\begin{aligned} U_1^{(\alpha)} &= 0 \\ U_2^{(\alpha)} &= \sin(m'_{s\alpha}\pi x_1/\tau_1)\cos(n'_\alpha \pi x_2/\tau_2) \end{aligned}\right\} \qquad \alpha = S_3 + 1, \quad S_3 + 2, \ldots, S_m \tag{3.38}$$

Here m_α, $n_{s\alpha}$, $m'_{s\alpha}$, and n'_α are all odd integers.

It is important to realize that a basic difference does exist between the motivation for using a corrective series in this section and in the previous section. In the previous section, the original choice of trial functions, Equations 3.18, is not pointwise complete and thus cannot yield solutions convergent to the right answers no matter how many terms are taken unless the corrective series is included. Here, however, the original choice of trial functions, Equations 3.37, is pointwise complete and will, by itself, yield solutions which do ultimately, albeit slowly, converge correctly. In other words, in this case the corrective series is merely added to accelerate convergence, while in the previous section it was vital to obtain meaningful answers at all.

Here, the result of mixing the corrective series with Equations 3.37 is very spectacular; convergence is improved by two orders of magnitude. However, certain cautions are necessary when using these over-complete expansions. In particular, as both Equations 3.37 and 3.38 form sets of functions which, at least normwise, are complete, the two sets taken together are not linearly independent as either set is extended to infinity. This fact can cause a breakdown in our technique with a finite number of trial functions if, when carrying out machine computations, a condition of near linear dependence causes the loss of too many significant figures in the eigenvalue subroutine.

In the course of performing numerical computation with a ten-place digital computer, a criterion has been observed for predicting in advance what combinations of the two series are likely to prove troublesomely degenerate: If two or more values $n_{s\alpha}$ occur in the corrective series for u_1 and if two or more values of $m'_{s\alpha}$ occur in the corrective series for u_2, we will nearly always encounter difficulties. On the other hand, if only one value of $n_{s\alpha}$ and $m'_{s\alpha}$ is present, difficulties are avoided in all combinations tested. This, in effect, means that it is best if the corrective series is just one-dimensional in character, as we stated without explaining in the previous section (see the explication of Equations 3.21).

The alert reader may have already pondered the possibility of taking Equations 3.38 for the main series and adding corrective terms of the form of Equations 3.37, instead of *vice versa* as we have been doing here. This should be possible, as Equations 3.38 are at least normwise complete and the additional terms from Equations 3.37 would make the overall expansion pointwise complete. In actuality, it is best not to do this because, while the expected correct convergence is found to occur, the rate is about one-third as fast as that resulting from taking Equations 3.37 as the main series. Whenever possible, it is best to start with a pointwise complete set of trial functions (which Equations 3.38 are not) and then to add over-complete terms if necessary to accelerate convergence. Starting with a trial function set which is not pointwise

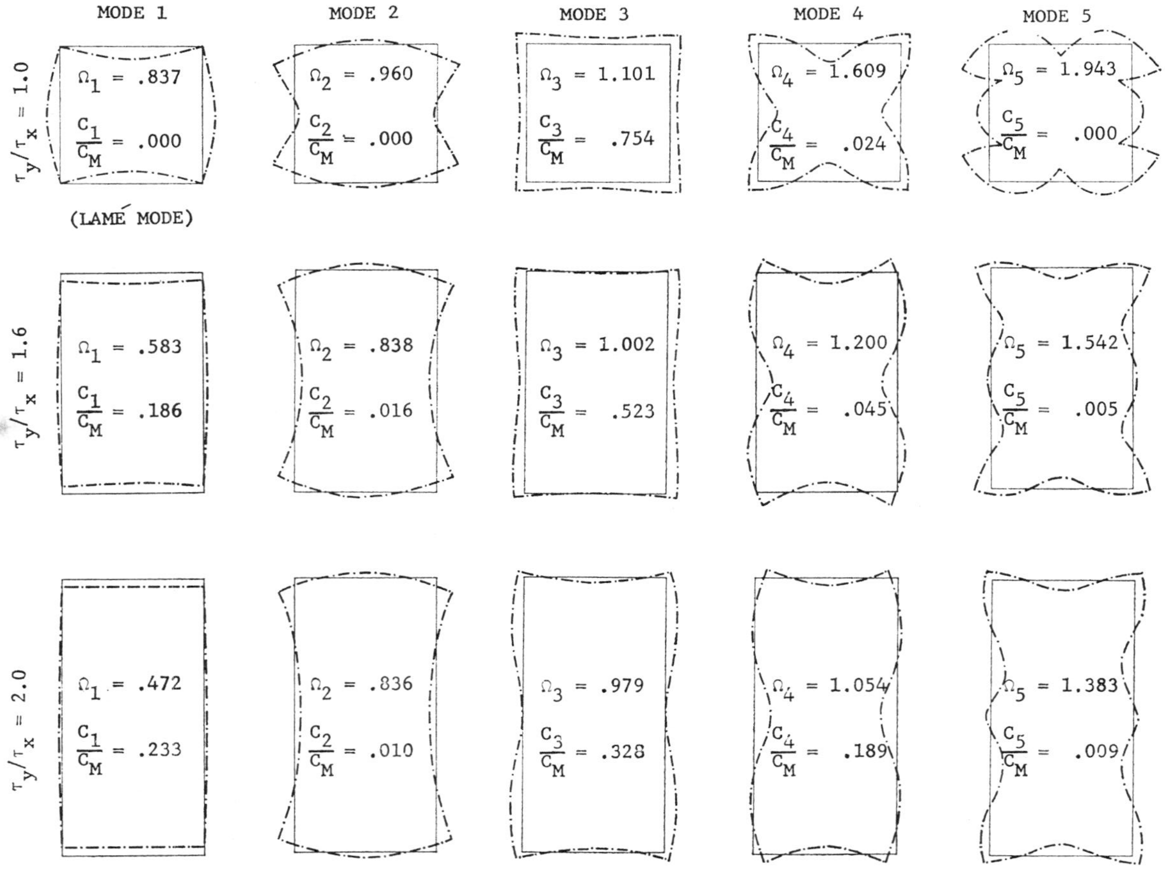

Figure 3.4. The lowest five dilation modes of a planarly isotropic rectangular plate for length-to-width ratios of 1.0, 1.6, and 2.0. (Based on the 40 trial functions of Equation 3.39, with $\gamma^E_{1122}/\gamma^E_{1111} = 0.30$; Ω_ν and C_ν/C_M are defined by Equations 3.42 and 3.56 where $C_\nu/C_M = C_\nu v$.)

complete and then patching it up, while a satisfactory approach, will yield slower convergence.

Recurring formulas for the elements of the P and H matrices of Equations 3.30 and 3.31, as they pertain to the trial functions of Equations 3.37 and 3.38, are tabulated elsewhere.[82] In the same report, a computer program is given for evaluating those recursion formulas and then solving the resulting characteristic value equation, Equation 3.32.

3.3.3. Numerical Results

Solutions for ω_v of the first 14 rectangular plate dilation modes may be obtained with 1 percent or better accuracy by taking 30 trial functions in the main series of Equations 3.37 and 10 corrective terms of Equation 3.38. Actual values assigned to the m's and n's which yield this accuracy are

$$(m_\alpha, n_{s\alpha}) = (n'_\alpha, m'_{s\alpha}) = (1,1), (3,1), (5,1), (7,1), (9,1)$$

$$\begin{aligned}(m_\alpha, n_{c\alpha}) = (n'_\alpha, m'_{c\alpha}) = {} & (1,0), (1,2), (1,4), (1,6), (1,8), (3,0),\\ & (3,2), (3,4), (3,6), (5,0), (5,2), (5,4),\\ & (7,0), (7,2), (9,0)\end{aligned} \tag{3.39}$$

Contour deformation patterns of the first five modes as determined in this way are illustrated in Figure 3.4 for length-to-width ratios

$$R = \tau_2/\tau_1 \tag{3.40}$$

of 1.0, 1.6, and 2.0. In these sketches, a planar Poisson's ratio

$$\sigma = \gamma^E_{1122}/\gamma^E_{1111} = -s^E_{12}/s^E_{11} \tag{3.41}$$

of 0.30 is assumed.

Let us now consider in detail the square plate case $R = 1$. Figure 3.5 shows the resonant frequency of the first 14 dilation modes versus σ for this configuration as computed from Equations 3.39. The values in this graph, and all succeeding numerical frequency values described in this section, are normalized into dimensionless form according to the rule

$$\Omega_v^2 = \frac{\tau_1^2 \omega_v^2 \rho}{\pi^2 \gamma^E_{1111}} \tag{3.42}$$

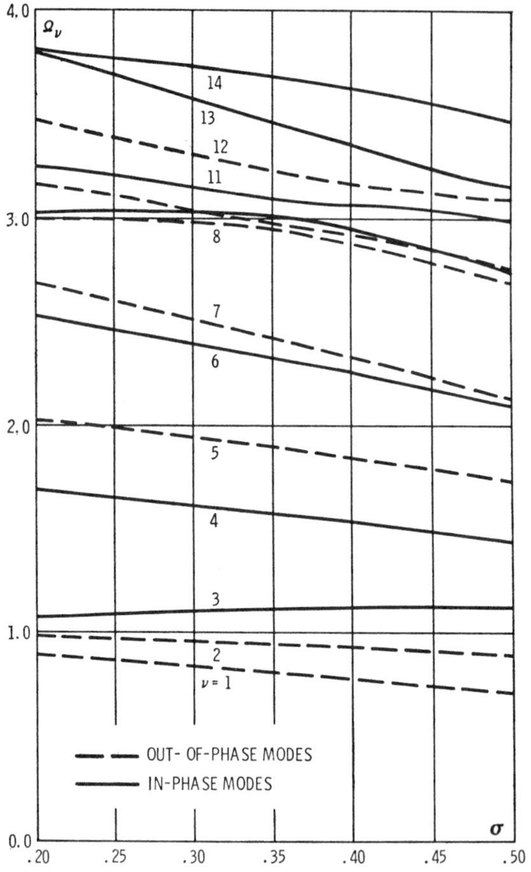

Figure 3.5. Frequency spectrum versus planar Poisson's ratio for the dilation modes of a planarly isotropic square plate.

In the case of a square plate, two types of dilation modes exist: those in which u_1 and u_2 are equal and in phase (modes 3, 4, 6, 10, 11, 13, 14, etc.) and those in which they are equal but out of phase (modes 1, 2, 5, 7, 8, 9, 12, etc.) Reference to the square plate deformation contours of Figure 3.4 and the spectral curves of Figure 3.5 illustrates this point.

For nonunity length-to-width ratios, no sharp distinction exists between in-phase and out-of-phase modes, as $u_1 \neq \pm u_2$. The normalized resonant frequencies for rectangular plates with $\sigma = 0.30$ are shown in Figure 3.6 as functions of R, the length-to-width ratio.

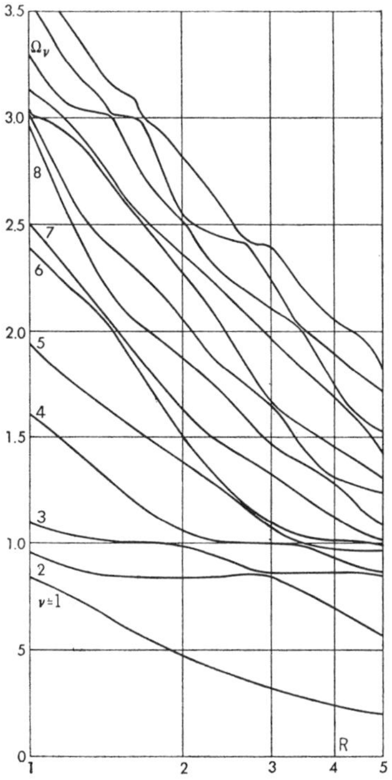

Figure 3.6. Frequency spectrum versus length-to-width ratio for the dilation modes of a planarly isotropic rectangular plate with $\sigma = 0.30$.

3.3.4. Comparison with Other Work on Rectangular Plates and Discussion of the Monotone Downward Convergence of Resonant Frequencies

It is interesting and instructive to compare resonant frequency values obtained from the present technique with those values obtained in other ways by various earlier authors.

The first work on this subject is that of Lamé[85] and Petrzilka,[86] in which mode 1 (see Figure 3.4), the so-called Lamé or equivolumal solution, was found for the square plate case. (Unlike other modal solutions to the elastic square or rectangular plate equations, this solution may be written exactly in closed form.) Petrzilka approached the differential equations involved by using separation of variables rather than the approximation techniques of more recent authors.

Ekstein[72] was the first to achieve reasonably accurate solutions for modes 2 and 3 (see Figure 3.4) of this family. He applied variational methods quite similar to ours and also did not satisfy either the differential equations or boundary conditions. Ekstein developed his theory for plates with an arbitrary length-to-width ratio but only performed numerical computations for square plates.

Earlier in this chapter, we observed that if we should choose trial functions which individually obey the boundary conditions exactly (even though this is not necessary) we will obtain much more accurate results with a given number of trial functions than otherwise. Actually, the question of whether or not the trial functions should obey the boundary conditions has another implication with respect to convergence: We may be concerned with the problem of obtaining lower bounds for the resonant frequencies.

It is well known that with the Rayleigh-Ritz variational technique we have been describing we always obtain approximate values for all the resonant frequencies which are too high (Rayleigh's theorem).[87,88] (This may not be true if the P or H matrices are not positive definite — a situation which can occur in the next chapter where piezoelectric effects are included in three-dimensional problems but not here in the two- or three-dimensional elastic case or two-dimensional piezoelectric plate case.) Consequently, it is not a simple matter to obtain lower bounds for the resonant frequencies when using variational procedures. However, there do exist methods of bounding these frequencies from below, most notably those of Temple and Weinstein.[87] It is important to note, though, that these methods require trial functions *that obey the boundary conditions individually*.

Mähly[89] treated the dilation-type square plate modes by adding a 10-term polynomial correction factor to Ekstein's solution for mode 3. This enabled him to match the boundary conditions and thus obtain a lower bound for ω_3. Other writers have, however, all considered this approach too cumbersome and have done without precise lower bounds. Mähly was the first to find that plate edges do not remain straight in square plate mode 3 (see Figure 3.4).

The most accurate existing work on square plate vibrations is that of Baerwald and Libove.[90] They used highly sophisticated trial functions that satisfied the differential equations and normal or dilational boundary conditions but violated the shear or tangential boundary conditions. Their work is believed to be accurate to six significant figures. However, it is limited by the restriction of being not easily applicable to rectangular plates of arbitrary length-to-width ratio. As the work of Baerwald and Libove is not widely available, the more important results have been reproduced in the IRE Standards.[91]

More recently, Onoe[92] has studied rectangular plates with variational techniques using trial functions that also obey the differential equations and normal boundary conditions. This general case is made more complicated than the square plate work of Baerwald and Libove by the loss of certain vibrational symmetries and the emergence of complex wavenumbers in the trial functions.

Additional work on this problem has been carried out by Medick and Pao.[93] Their technique consists of expanding the displacements across the smaller plate dimension in terms of Legendre polynomials and across the larger dimension in sines and cosines. The sine and cosine wavenumbers may become complex here also when made to satisfy the condition that the overall functions obey the wave equation approximately. Linear combinations of these sine-cosine-Legendre trial functions are then compounded so as to meet the boundary conditions approximately. It is noteworthy that certain of the parameters in this work are adjusted in a nonvariational manner that does not guarantee the monotone downward convergence of the resonant frequency approximations.

Most recently, rectangular elastic plates have been studied by Lloyd and Redwood with finite difference methods.[83] These methods also do not necessarily yield monotone downward convergence.[87]

Significant experimental results on rectangular elastic plates have been reported by Lloyd and Redwood,[83] Onoe,[92] Onoe and Pao,[94] EerNisse,[95] and Holland.[81,82]

In Table 3.2, we present a comparison of resonant frequency values for modes 2, 3, and 4 in the square plate case. Shown here are values obtained by the authors mentioned earlier and values obtained from the variational procedure described in this section with the 40 trial functions of Equations 3.39. Also given in Table 3.2 are the results found when the present procedure is used but with 86 rather than 40 trial functions. As mentioned previously, the work of Baerwald and Libove is thought to be the most accurate and should be regarded as the standard. Their values for Ω_3 and Ω_4 are almost certainly correct to the four decimal places given in Table 3.2.

Use of 86 rather than 40 trial functions produces slightly closer correspondence between our solutions and those of Baerwald and Libove. However, we have found empirically that with the variational procedures described in this chapter, computer time increases as the 2.5 power of the number of trial functions. Consequently, the slight increase in accuracy obtained with 86 rather than 40 trial functions is not deemed adequate recompense for the increased computer time required.

Table 3.2. Comparison of Normalized Resonant Frequencies of Dilation Modes 2, 3, and 4: Present Technique Versus Other Techniques (Based on a Square Plate with $\sigma = 0.32$).

	Ω_2	Ω_3	Ω_4
Ekstein[72]	1.0000	1.1221	†
Mähly (solution)[89]	†	1.1043	†
Mähly (lower bound)[89]	†	1.0964	†
Baerwald and Libove[90,91]	†	1.1033	1.5891
Onoe (theoretical)[92]	isotropic case not considered		
Medick and Pao (theoretical)[93]	0.844	1.131	1.725
Lloyd and Redwood (theoretical)[83]	0.935	1.100	1.528
Onoe and Pao (experimental)[93,94]	0.905	1.066	1.539
EerNisse (experimental)[95]	†	1.116	1.556
Holland (experimental)[81,82]	†	*	1.616
Present technique			
30 terms in main series and 10 corrective terms	0.9548	1.1055	1.5929
72 terms in main series and 14 corrective terms	0.9509	1.1047	1.5916

* In this work, σ was determined experimentally by matching the theoretical and experimental values of Ω_3.

† Not determined.

We can observe from Table 3.2 that the results of the present technique with just 40 trial functions come considerably closer than any others to agreeing with those of Baerwald and Libove. (We should emphasize again that the work of Baerwald and Libove is an excellent standard but not a panacea, as it is only applicable to square plates and not to general rectangular plates.)

Two of our trial functions from Equations 3.39,

$$(m_\alpha, n_{s\alpha}) = (1,1)$$
$$(n'_\alpha, m'_{s\alpha}) = (1,1) \qquad (3.43)$$

may be combined to yield the Lamé solution, mode 1, exactly. Consequently, our procedure will give back this solution (and resonant frequency) exactly. This is also true of the other theoretical procedures considered in Table 3.2, excepting that of Lloyd and Redwood. For this reason, comparative data on mode 1 has been omitted from Table 3.2.

3.3.5. Determination of the Equivalent Circuit Characteristics in the Piezoelectric Case

The resonant modes and frequencies computed in this section pertain to any rectangular plate isotropic in its own plane, be it piezoelectric or not. However, now let us restrict our attention to the piezoelectric case and consider the plate to be ferroelectric ceramic, poled in the x_3-direction, and completely electroded on the major faces. In this case, the modes and resonant frequencies computed here may be used with the theory of Section 2.3 to obtain the admittance and equivalent circuit characterizing the plate electrically.

To be precise, Equations 2.62 and 2.71 indicate that the electrical admittance of a plate such as this is

$$Y = j\omega C_0 + j\omega \sum_{\nu} \frac{\left(\iint_A \tilde{e}_{3ij} u_{i,j}^{(\nu)} \, dA\right)^2}{\omega_\nu^2 - \omega^2} \tag{3.44}$$

In this equation, $u_{i,j}^{(\nu)}$ is the νth eigenmode of Equation 3.32, normalized according to Equation 2.63, or equivalently, according to Equation 3.34.

Equation 3.44 may be more conveniently expressed by using the divergence theorem to convert the area integral into a contour integral around the edge of the plate,

$$Y = j\omega C_0 + j\omega \sum_{\nu} \frac{\left(\oint_{\partial A} \tilde{e}_{3ij} u_i^{(\nu)} N_j \, ds\right)^2}{\omega_\nu^2 - \omega^2} \tag{3.45}$$

Here ∂A represents the boundary contour of the plate, and N_j is the outward pointing unit normal vector of ∂A.

As Equation 3.45 describes a structure with a single terminal pair, we can represent it by a much simpler equivalent circuit than those given in Section 2.3. In particular, this admittance may be pictorially described at all ω by the infinite parallel combination of LC-series resonant circuits shown in Figure 3.7.

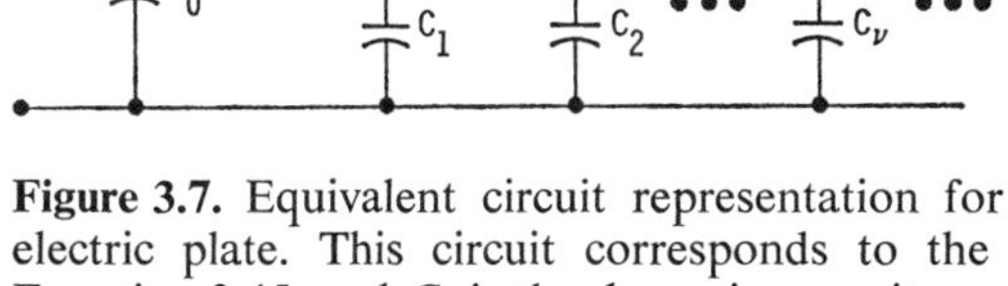

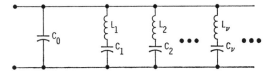

Figure 3.7. Equivalent circuit representation for a rectangular piezoelectric plate. This circuit corresponds to the admittance given by Equation 3.45, and C_ν is the dynamic capacitance of mode ν.

Let us now reflect on what values one must assign the circuit elements in Figure 3.7 in order to permit quantitative representation of Equation 3.45. First of all, the νth LC-series combination of Figure 3.7 represents mode ν and thus must be resonant at the νth resonance of the plate,

$$\omega_\nu^2 = 1/L_\nu C_\nu \tag{3.46}$$

In addition, we can write the admittance of the equivalent circuit in Figure 3.7 as

$$Y = j\omega C_0 + j\omega \sum_\nu \frac{C_\nu \omega_\nu^2}{\omega_\nu^2 - \omega^2} \tag{3.47}$$

Comparison of Equations 3.47 and 3.45 then indicates that the equivalent dynamic capacitance associated with mode ν in Figure 3.7 must be

$$C_\nu = \frac{1}{\omega_\nu^2}\left(\oint_{\partial A} \tilde{e}_{3ij} u_i^{(\nu)} N_j \, ds\right)^2 \tag{3.48}$$

For modes characterized by a large C_ν, the plate admittance in the vicinity of resonance ω_ν will be high. This, in turn, is usually considered to mean that the resonance at ω_ν is strong. In other words, the size of C_ν may be interpreted as a measure of the piezoelectric strength of mode ν.

In Equations 3.45 and 3.47, the C_0 represents the capacitance of the plate at high frequency, *i.e.*, when it is inertially clamped in its own plane, Equation 2.70,

$$C_0 = \epsilon_{33}^{S_2} \tau_1 \tau_2 / \tau \tag{3.49}$$

Here $\epsilon_{33}^{S_2}$ is given by Equation 2.47,

$$\epsilon_{33}^{S_2} = \epsilon_{33}^{T} - \tilde{e}_{3ij} \tilde{e}_{3kl} s_{ijkl}^{E} \tag{3.50}$$

On the other hand, the very low-frequency capacitance of the plate is

$$C_T = \epsilon_{33}^{T} \tau_1 \tau_2 / \tau \tag{3.51}$$

Figure 3.7 indicates that the difference between C_T and C_0 must be the sum of all the dynamic capacitances C_ν,

$$C_T - C_0 = \sum_\nu C_\nu = \tilde{e}_{3ij}\tilde{e}_{3kl}s^E_{ijkl}\tau_1\tau_2/\tau \tag{3.52}$$

since all the L_ν look like short circuits at very low frequency.

Consequently, we may describe the relative piezoelectric strengths of each mode by what we shall call the normalized dynamic capacitance of that mode,

$$C_{\nu N} = C_\nu/(C_T - C_0) \tag{3.53}$$

This quantity is not only dimensionless but proves also to be independent of the actual piezoelectric tensor for the plate. The sum of all the $C_{\nu N}$ must, of course, be 1 because of Equations 3.52 and 3.53.

If we note that for a ferroelectric ceramic $\tilde{e}_{311} = \tilde{e}_{322}$ are the only nonzero components of $\tilde{e}_{3ij}$, we can express $C_{\nu N}$ as

$$C_{\nu N} = \frac{(1+\sigma)}{2\Omega_\nu^2}\frac{\rho\tau}{R\pi^2}\left(\oint_{\partial A} u_N^{(\nu)}\, ds\right)^2 \tag{3.54}$$

In this equation, $u_N^{(\nu)}$ is the component of $u_i^{(\nu)}$ along N_i and use has been made of Equations 3.42, 2.49, 2.45, 1.30, and 3.41. It is apparent from Equation 3.54 that $C_{\nu N}$ is indeed independent of the piezoelectric coefficient $\tilde{e}_{311}$.

In point of fact, $C_{\nu N}$ depends on only two parameters R and σ. To see that this is so, we re-express the normalized displacement pattern $u_N^{(\nu)}$ as

$$\psi_N^{(\nu)} = \frac{\tau_1}{\pi}\sqrt{\rho\tau}\, u_N^{(\nu)} \tag{3.55}$$

where, according to Equation 2.63, the $\psi_N^{(\nu)}$ is completely dimensionless. This definition reduces Equation 3.54 to

$$C_{\nu N} = \frac{(1+\sigma)}{2\Omega_\nu^2 R}\left(\oint_{\partial A} \psi_N^{(\nu)}\, d\sigma\right)^2 \tag{3.56}$$

where $d\sigma = ds/\tau_1$. All quantities appearing in this equation are dimensionless and depend only on R and σ.*

Equation 3.56 has been evaluated in a number of situations by means of the $\psi_N^{(\nu)}$ computed from Equations 3.33 and 3.55. Results are plotted

* The fact that $\psi_N^{(\nu)}$ and Ω_ν depend only on R and σ is proved in Reference 82.

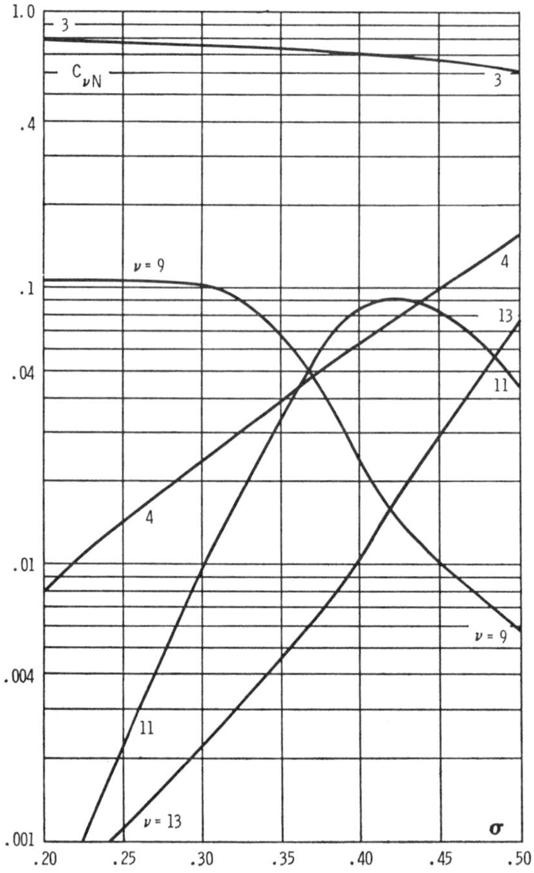

Figure 3.8. Normalized dynamic capacitances versus planar Poisson's ratio for the dilation modes of a planarly isotropic square plate.

versus σ for the square plate case ($R = 1$) in Figure 3.8. For square plates where x_1 and x_2 modal displacement components are either in phase ($u_1 = u_2$) or out of phase ($u_1 = -u_2$), only the in-phase modes have nonzero $C_{\nu N}$. In the case of the out-of-phase modes, any contribution to $C_{\nu N}$ by u_1 is exactly cancelled by a corresponding contribution from u_2.

The values of $C_{\nu N}$ for a rectangular plate with $\sigma = 0.30$ are shown in Figure 3.9 when $1 \leq R \leq 5$. Because of the symmetry of this geometry and material, these curves are invariant to the transformation

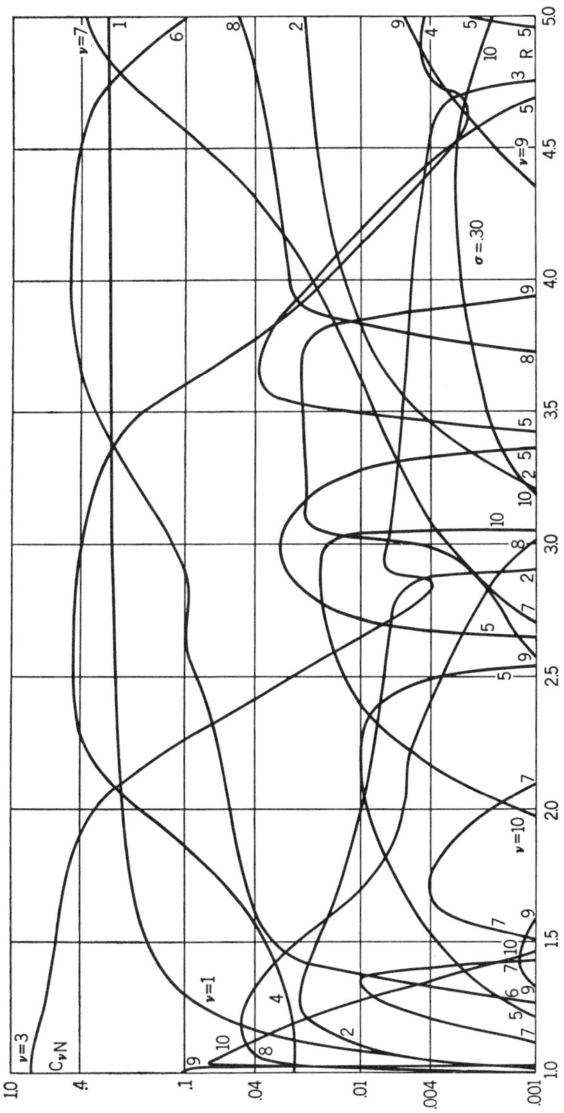

Figure 3.9. Normalized dynamic capacitances versus length-to-width ratio for the dilation modes of a rectangular plate. (Based on $\sigma = 0.30$.)

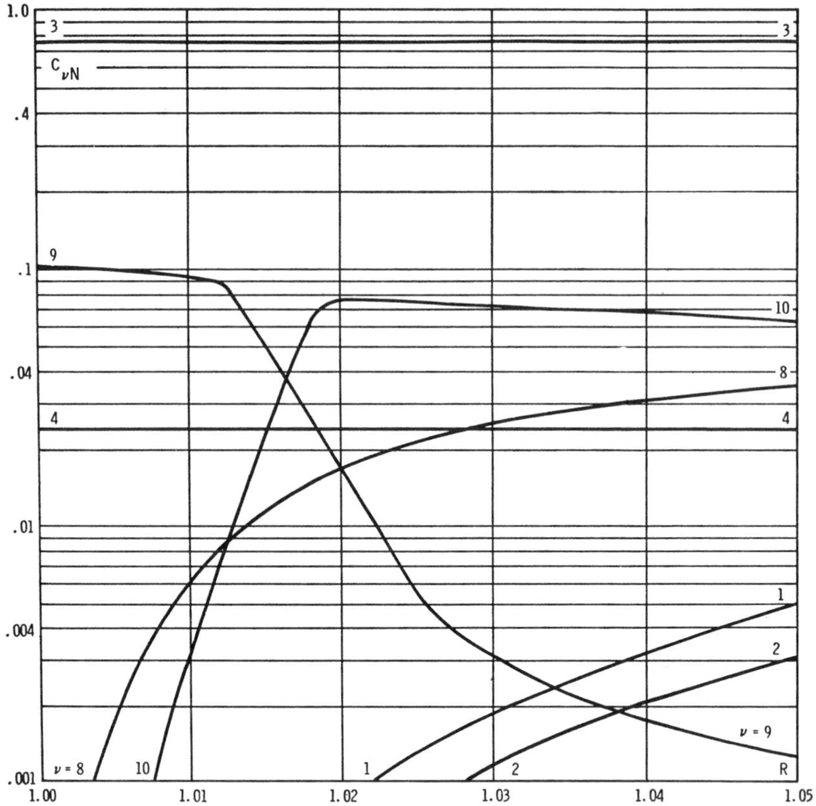

Figure 3.10. Expanded plot of the normalized dynamic capacitances versus length-to-width ratio near $R = 1$ for the dilation modes of a planarly isotropic rectangular plate. (Based on $\sigma = 0.30$.)

$R \to 1/R$ (*i.e.*, 90° rotation of the plate in the x_1x_2-plane). Consequently, the curves in Figure 3.9 are equally applicable when $1 \geq R \geq \frac{1}{5}$.

For nonunity length-to-width ratios, there is no clear distinction between in-phase and out-of-phase modes. Thus the $C_{\nu N}$ of modes which are out of phase at $R = 1$ change rather suddenly from their zero value at $R = 1$ to nonzero values when R is near but not equal to 1. Figure 3.10 shows an expanded plot of the $C_{\nu N}$ near $R = 1$ where this transition takes place.

4 Variational Treatment of Piezoelectric Problems

In Chapter 3 we discussed the variational approach for analyzing mechanically free resonant elastic structures. Piezoelectric effects were considered only in thin, fully electroded contour-extensional plate resonators, where the introduction of piezoelectricity does not perturb the elastic equations of motion.

In general three-dimensional problems, though, piezoelectric effects will alter the elastic equations of motion and must be explicitly included in those equations. Moreover, we may need to treat mechanical boundary conditions other than the simple free case. Also, when piezoelectric phenomena are present, we must consider not merely mechanical boundary conditions but electrical ones as well.

Recently the classical variational techniques of elasticity have been extended to handle these more general problems where volume piezoelectric effects and complicated boundary conditions are of concern.[96,97,98] This new work will be our subject for this chapter.

We shall discuss, for instance, piezoelectric structures such as that shown in Figure 4.1, where the electrode shape and number, sample shape, and external force distribution are all arbitrary. Under linear conditions, the most concise means of describing such a configuration is by an electromechanical admittance or impedance matrix. In Chapter 2 we offered descriptions of this type for the special two-dimensional situation. Here we shall demonstrate that we can variationally derive similar formulas for these matrices in the general three-dimensional case, provided we have knowledge of the natural modes of the structure in question.

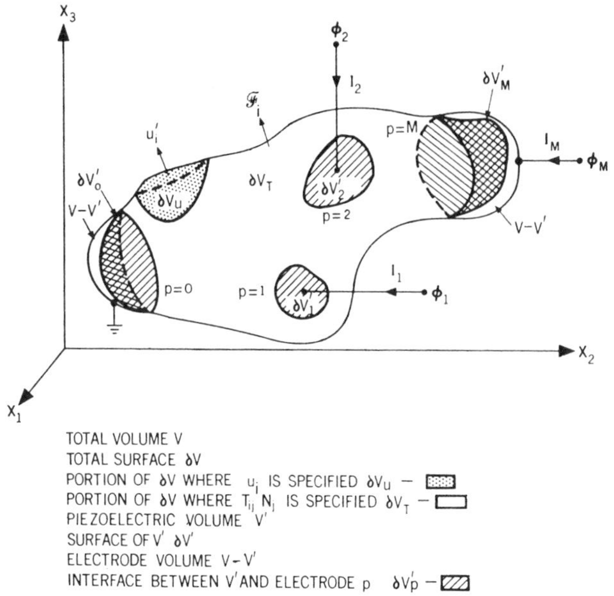

Figure 4.1. A three-dimensional piezoelectric structure with arbitrary shape, electrode configuration, and external force distribution.

These natural modes, however, obey equations of motion that almost always cannot be solved exactly in three-dimensional cases. In Chapter 3, we demonstrated classical variational techniques for obtaining approximate solutions to similar nonpiezoelectric equations that could not be solved exactly. The new methods described here extend this variational approach in a way that allows approximate evaluation of natural modes in the presence of piezoelectricity. Consequently, it is possible with the work reported in this chapter, first, to find the natural modes of arbitrary electroelastic structures such as we see in Figure 4.1 and, second, to use those modes for evaluating the structure's admittance or impedance matrix.

At the end of this chapter, we shall illustrate the application of these techniques to rectangular piezoelectric parallelepipeds and thick piezoelectric disks.

4.1. Waves in Unbounded Media; Four-Dimensional Formulation

4.1.1. Waves in Unbounded Media

To introduce the application of variational calculus in piezoelectricity, let us consider a fictitious problem in which all space is filled with piezoelectric medium. This medium need not be homogeneous, but we will assume linear sinusoidal conditions prevail. Moreover, we will assume any motions which are present vanish as one retreats infinitely far from the origin, so that boundary effects at infinity may be neglected.

In this case, the constituent relations for a material of arbitrary crystallographic symmetry are best written as in Equation 1.28,

$$\begin{aligned} T_{ij} &= c^E_{ijkl} S_{kl} - e_{mij} E_m \\ D_n &= e_{nkl} S_{kl} + \epsilon^S_{mn} E_m \end{aligned} \tag{4.1}$$

When these relations are expressed with particle displacement u_k and electric potential ϕ as independent variables, we get

$$\begin{aligned} T_{ij} &= c^E_{ijkl} u_{k,l} + e_{mij} \phi_{,m} \\ D_n &= e_{nkl} u_{k,l} - \epsilon^S_{mn} \phi_{,m} \end{aligned} \tag{4.2}$$

Commas again denote differentiation, which is understood to be covariant in cases of non-Cartesian coordinates.

Conservation of momentum is now expressed in differential form as

$$T_{ij,j} = (c^E_{ijkl} u_{k,l} + e_{mij} \phi_{,m})_{,j} = -\rho\omega^2 u_i \tag{4.3}$$

Additionally, for insulating materials D_n must be divergenceless,

$$D_{n,n} = (e_{nkl} u_{k,l} - \epsilon^S_{mn} \phi_{,m})_{,n} = 0 \tag{4.4}$$

Again, the problem of describing this situation variationally consists of finding some Lagrangian expression with a first variation which is zero *only if* Equations 4.3 and 4.4 hold.[73] The following expression possesses this property:

$$\begin{aligned} L &= \iiint_\infty \tfrac{1}{2}(T_{ij} S_{ij} - D_m E_m - \rho\omega^2 u_i u_i)\, dV \\ &= \iiint_\infty \tfrac{1}{2}(u_{i,j} c^E_{ijkl} u_{k,l} + 2\phi_{,m} e_{mkl} u_{k,l} - \phi_{,m} \epsilon^S_{mn} \phi_{,n} - \rho\omega^2 u_i u_i)\, dV \end{aligned} \tag{4.5}$$

In particular, if arbitrary independent variations are assigned to u_i, $u_{i,j}$, ϕ, and $\phi_{,m}$, we find the first variation in L is

$$\begin{aligned} \delta L = &\iiint_\infty \{[(-u_{i,j} c^E_{ijkl} - \phi_{,m} e_{mkl})_{,l} - \rho\omega^2 u_k]\, \delta u_k \\ &+ [(-e_{mkl} u_{k,l} + \epsilon^S_{mn} \phi_{,n})_{,m}]\, \delta\phi\}\, dV \\ &+ \oiint_\infty \{(u_{i,j} c^E_{ijkl} + \phi_{,m} e_{mkl})\, \delta u_k N_l \\ &+ (e_{mkl} u_{k,l} - \epsilon^S_{mn} \phi_{,n})\, \delta\phi N_m\}\, dA \end{aligned} \tag{4.6}$$

As we have postulated here that no motions take place at infinity, the surface integral in Equation 4.6 contributes nothing. This, in fact, is true even if the variations δu_k and $\delta\phi$ do not obey the boundary condition of vanishing at infinity, just so long as the two surface integrand factors in parentheses vanish there.

Consequently, as δu_k and $\delta\phi$ are specified to be arbitrary, this δL can vanish irrespective of δu_k and $\delta\phi$ *only if* the two bracketed terms in the volume integrand are zero at all finite points. However, this requirement implies Equations 4.3 and 4.4. Hence Equation 4.5 is an appropriate Lagrangian for piezoelectric infinite medium problems.

It is curious to observe that the potential energy term in the present Lagrangian is not positive definite. While this fact has no fatal consequences in variational work, it does imply that Rayleigh-Ritz procedures based on Equation 4.5 will not necessarily exhibit the usual monotone covergence. This phenomenon is in direct opposition to the situation found in Chapter 3 where purely elastic problems were presented.

4.1.2. *Four-Dimensional Formulation**

Let us define the four-space coordinate system

$$x_p = (x_i, t) \tag{4.7}$$

and the four-vector

$$u_p = (u_i, \phi) \tag{4.8}$$

where subscripts p, q, r, s will be assumed to run 1 to 4. Also define the second-rank four-tensor

$$\rho_{pq} = \begin{cases} \rho\delta_{pq} & p, q = 1, 2, 3 \\ 0 & p, q = 4 \end{cases} \tag{4.9}$$

and the fourth-rank four-tensor M_{pqrs}, where

$$\begin{aligned} &M_{ijkl} = c^E_{ijkl}, \qquad M_{4jkl} = e_{jkl}, \qquad M_{ijk4} = e_{kij}, \\ &M_{4jk4} = -\epsilon^S_{jk}, \qquad M_{p44s} = -\rho_{ps} \end{aligned} \tag{4.10}$$

and all other components of $M_{pqrs} = 0$. Then one can express the equations of motion, Equations 4.3 and 4.4, simultaneously as one

* This subsection may be omitted without loss of continuity.

four-space equation

$$(u_{p,q}M_{pqrs})_{,r} = 0 \tag{4.11}$$

(Try it!)

While Equation 4.11 is superficially reminiscent of certain relativistic conservation equations, it apparently lacks any real significance in that respect. For Equation 4.11 to be actually useful in problem solving, the speed of sound in a piezoelectric material would need to be invariant to transformation into moving coordinate systems. However, not only is that requirement unfulfilled but the very concept of a moving coordinate system is irrelevant to our present work.

Still, the idea introduced by Equation 4.8 that ϕ may be regarded as a fourth component of u_i is a rather interesting one. From a mathematical viewpoint, at least, it is simplifying and consistent to consider ϕ from this attitude.

4.2. Lagrangian Formulation for Determination of Admittance and Impedance Matrices of Three-Dimensional Piezoelectric Bodies

4.2.1. The Lagrangian or Stationary Expression

Let us now take up formal consideration of the arbitrary piezoelectric sample shown in Figure 4.1. We shall indicate the volume of the entire structure by V and the piezoelectric volume (the insulating region occupied by the sample exclusive of the electrodes) by V'. Thus the electrodes occupy $V - V'$ (see Figure 4.1). If we permit a force per unit volume f_i to act within V, then the momentum conservation equation for a differential volume element is

$$T_{ij\,j} + \rho\omega^2 u_i = (c^E_{ijkl}u_{k,l} + e_{nij}\phi_{,n})_{,j} + \rho\omega^2 u_i = -f_i \qquad \text{in } V \tag{4.12}$$

where we again assume an $e^{j\omega t}$ time dependence.

Inside the piezoelectric region V', no free charge is permitted. Thus the electric displacement must have no divergence there,

$$D_{m,m} = (e_{mkl}u_{k,l} - \epsilon^S_{mn}\phi_{,n})_{,m} = 0 \qquad \text{in } V' \tag{4.13}$$

Let us designate the surface of V as ∂V and that of V' as $\partial V'$. For a structure such as this the boundary conditions on ∂V and $\partial V'$ are quite complex. Mechanically, the most general case is that in which a linear combination of u_i and the normal component of T_{ij}, the N_jT_{ij}, is prescribed on ∂V. However, we shall restrict our attention to the special case in which either u_i or $N_j T_{ij}$ is specified at every point on ∂V,

$$u_i = u_i' \qquad \text{on } \partial V_u \tag{4.14}$$

and

$$N_j T_{ij} = N_j(c_{ijkl}^E u_{k,l} + e_{nij}\phi_{,n}) = \mathscr{F}_i \qquad \text{on } \partial V_T \tag{4.15}$$

where

$$\partial V_u + \partial V_T = \partial V \tag{4.16}$$

Here u_i' and $\mathscr{F}_i$ are the prescribed quantities.

Electrically, one must specify the free surface charge density λ and the external electric displacement on the unelectroded portion of the surface of V', which we shall designate $\partial V_\lambda'$. It is, however, usually sufficient (and always convenient) to neglect external fields. If this is done, the free surface charge density on $\partial V_\lambda'$ is equal to minus the outward normal component of the electric displacement,

$$-N_m D_m = -N_m(e_{mkl}u_{k,l} - \epsilon_{mn}^S \phi_{,n}) = \lambda \qquad \text{on } \partial V_\lambda' \tag{4.17}$$

At the interface between V' and an electrode, we may either specify the potential or the total free charge. Let us assume the first condition is met at $M + 1$ electrodes indexed by p or q with interface surfaces $\partial V_p'$, and the second is met at M' electrodes indexed by r or s with interface surfaces $\partial V_r'$. We then have

$$\phi = \phi_p \qquad \text{on } \partial V_p' \qquad p = 0, 1, \ldots, M \tag{4.18}$$

and

$$-\iint_{\partial V_r'} N_m(e_{mkl}u_{k,l} - \epsilon_{mn}^S \phi_{,n})\, dA = Q_r \qquad r = 1, 2, \ldots, M' \tag{4.19}$$

where

$$\partial V_\lambda' + \sum_{p=0}^{M} \partial V_p' + \sum_{r=1}^{M'} \partial V_r' = \partial V' \tag{4.20}$$

and where ϕ_p and Q_r represent the specified values. In the case of the M' electrodes at which Q_r is specified, the auxiliary condition is required that ϕ be a constant over $\partial V_r'$,

$$\phi = \phi_r \quad \text{on} \quad \partial V_r' \qquad r = 1, 2, \ldots, M' \tag{4.21}$$

Unlike the ϕ_p, these constants ϕ_r are not known quantities.

When working with electric potentials, it is always necessary to define a reference level. This will be done here by considering the potential at electrode $p = 0$ to be ground, $\phi_0 = 0$ (see Figure 4.1).

The problem of formulating this general situation variationally is again just a matter of finding a Lagrangian with a first variation which is zero *only if* Equations 4.12 to 4.21 hold.[73] Primarily by trial and error, the following expression was found which meets this requirement:

$$\begin{aligned} L = & \iiint_V \tfrac{1}{2}(u_{i,j} c_{ijkl}^E u_{k,l} - \rho\omega^2 u_i u_i)\, dV - \iiint_V u_i f_i \, dV \\ & + \iiint_{V'} \tfrac{1}{2}(2\phi_{,m} e_{mkl} u_{k,l} - \phi_{,m} \epsilon_{mn}^S \phi_{,n})\, dV - \iint_{\partial V_T} \mathscr{F}_i u_i \, dA \\ & - \iint_{\partial V_u} (u_i - u_i') N_j (c_{ijkl}^E u_{k,l} + e_{nij}\phi_{,n})\, dA + \iint_{\partial V_\lambda'} \lambda \phi \, dA \\ & - \sum_{p=0}^{M} \iint_{\partial V_p'} (\phi - \phi_p) N_m (e_{mkl} u_{k,l} - \epsilon_{mn}^S \phi_{,n})\, dA \\ & - \sum_{r=1}^{M'} \iint_{\partial V_r'} (\phi - \phi_r) N_m (e_{mkl} u_{k,l} - \epsilon_{mn}^S \phi_{,n})\, dA + \sum_{r=1}^{M'} \phi_r Q_r \end{aligned} \tag{4.22}$$

The first variation δL of this quantity is computed analogously to Equations 3.7 and 4.6. First assign an arbitrary independent infinitesimal variation to each of the unknowns, u_i, ϕ, ϕ_r, and to the derivatives of u_i and ϕ but not to the known quantities $\omega, f_i, \mathscr{F}_i, u_i', \lambda, Q_r$, and ϕ_p. Then evaluate the corresponding first-order change in L. This change is the desired δL. The variations in u_i and ϕ may be any infinitesimal functions possessing second derivatives. They need not obey the differential equations or boundary conditions. The variations in each of the ϕ_r, on the other hand, are not functions of position since the ϕ_r pertain to the potentials at electrodes r, and these are a constant over the $\partial V_r'$ (see Equation 4.21).

Evaluation of δL in this manner and use of the divergence theorem lead to the result,

$$
\begin{aligned}
\delta L = &-\iiint_V \delta u_i[(c^E_{ijkl}u_{k,l} + e_{nij}\phi_{,n})_{,j} + \rho\omega^2 u_i + f_i]\, dV \\
&-\iiint_{V'} \delta\phi[(e_{mkl}u_{k,l} - \epsilon^S_{mn}\phi_{,n})_{,m}]\, dV \\
&-\iint_{\partial V_u} [u_i - u_i']N_j(c^E_{ijkl}\delta\{u_{k,l}\} + e_{nij}\delta\{\phi_{,n}\})\, dA \\
&-\sum_{p=0}^{M} \iint_{\partial V_p'} [\phi - \phi_p]N_m(e_{mkl}\delta\{u_{k,l}\} - \epsilon^S_{mn}\delta\{\phi_{,n}\})\, dA \\
&-\sum_{r=1}^{M'} \iint_{\partial V_r'} [\phi - \phi_r]N_m(e_{mkl}\delta\{u_{k,l}\} - \epsilon^S_{mn}\delta\{\phi_{,n}\})\, dA \\
&+\iint_{\partial V_T} \delta u_i[N_j(c^E_{ijkl}u_{k,l} + \phi_{,n}e_{nij}) - \mathscr{F}_i]\, dA \\
&+\iint_{\partial V_\lambda'} \delta\phi[N_m(e_{mkl}u_{k,l} - \epsilon^S_{mn}\phi_{,n}) + \lambda]\, dA \\
&+\sum_{r=1}^{M'} \delta\phi_r\left[\iint_{\partial V_r'} N_m(e_{mkl}u_{k,l} - \epsilon^S_{mn}\phi_{,n})\, dA + Q_r\right]
\end{aligned} \tag{4.23}
$$

The $\delta\phi$, $\delta\phi_r$, δu_i, $\delta\{\phi_{,n}\}$, $\delta\{u_{k,l}\}$ have all been defined as arbitrary. Consequently, we can set any one of them equal to a Gaussian pulse or delta function acting at any point included in the domains of integration or summation in Equation 4.23 and set the others equal to zero or functions of zero measure. With this freedom of choice in $\delta\phi$, $\delta\phi_r$, and δu_j, the variation in L will vanish irrespective of $\delta\phi$, $\delta\phi_r$, and δu_j *only if* each of the eight bracketed terms is zero at every point in those domains. However, this requirement will be met *only if* Equations 4.12 through 4.21 hold. Hence, the Lagrangian of Equation 4.22 is appropriate for the problem of Figure 4.1.

4.2.2. The Admittance Matrix

In most problems of practical interest the majority of the surface terms in the Lagrangian of Equation 4.22 may be set equal to zero.

Consider, for example, the special case in which the external mechanical force is specified on the entire boundary and the displacement is specified nowhere. Then Equations 4.14 and 4.16 become

$$\begin{aligned} \partial V_u &= 0 \\ \partial V_T &= \partial V \end{aligned} \tag{4.24}$$

Similar restrictions on the electrical boundary conditions may often exist. For example, assume that there is no free charge on the unelectroded portion of the surface and that the potential rather than the total charge is specified on all the electrodes,

$$\begin{aligned} M' &= 0 \\ \lambda &= 0 \qquad \text{on } \partial V_\lambda' \end{aligned} \tag{4.25}$$

Under the limitations of Equations 4.24 and 4.25, it is most convenient to describe the linear response of the system in Figure 4.1 by its electromechanical admittance matrix. In particular, the Lagrangian of Equation 4.22 now reduces to

$$\begin{aligned} L = &\iiint_V \tfrac{1}{2}(u_{i,j}c^E_{ijkl}u_{k,l} - \rho\omega^2 u_i u_i)\, dV - \iiint_V u_i f_i \, dV \\ &+ \iiint_{V'} \tfrac{1}{2}(2\phi_{,m}e_{mkl}u_{k,l} - \phi_{,m}\epsilon^S_{mn}\phi_{,n})\, dV - \iint_{\partial V_T} \mathscr{F}_i u_i \, dA \\ &- \sum_{p=0}^{M} \iint_{\partial V_p'} (\phi - \phi_p)N_m(e_{mkl}u_{k,l} - \epsilon^S_{mn}\phi_{,n})\, dA \end{aligned} \tag{4.26}$$

To determine the admittance matrix in this case, we first evaluate the particle displacement and potential patterns u_i and ϕ. This is done by expansion in terms of the normal modes of the system. Let $u_i^{(\nu)}$ and $\phi^{(\nu)}$ be the νth particle displacement and potential eigenmode solutions of Equations 4.12 to 4.21 with f_i, ∂V_u, $\mathscr{F}_i$, λ, ϕ_p, and M' all zero. In other words, these eigenmodes represent the homogeneous or unforced solutions of the body shown in Figure 4.1 with all the electrodes short circuited. In Section 4.3, we shall indicate how these eigenmodes may be evaluated variationally. For the present, we shall simply assume they are accessible.

We then utilize the fact that u_i and ϕ may be expanded exactly in the form

$$
\begin{aligned}
u_i &= \sum_{\nu} A^{(\nu)} u_i^{(\nu)} \\
\phi &= \sum_{\nu} A^{(\nu)} \phi^{(\nu)} + \phi''
\end{aligned}
\tag{4.27}
$$

where the $A^{(\nu)}$ are generalized Fourier coefficients to be evaluated variationally.

The ϕ'' appearing in Equation 4.27 may be interpreted in several ways. For instance, while it is generally true that the $u_i^{(\nu)}$ form a complete set of functions, this is not true of the $\phi^{(\nu)}$. Consequently, it is necessary to add a correction term to represent the projection of ϕ into the function subspace which the $\phi^{(\nu)}$ do not span. On more physical grounds, we can argue that as frequency goes to infinity, inertial clamping constrains particle displacement (and hence all the $A^{(\nu)}$) to zero for any finite drive. However, the potential distribution is not similarly constrained. Thus, we can regard the ϕ'' as the infinite frequency potential or, alternatively, the potential with all mechanical motion prevented. The necessity for this ϕ'' term in such expansions was apparently first recognized by Lewis.[35]

The initial step in the evaluation of the $A^{(\nu)}$ coefficients is to substitute Equations 4.27 into 4.26. However, before we do this, a few comments should be made.

First, each of the short-circuit potential eigenmode solutions is zero by definition on all the electrodes. Consequently, Equations 4.27 will yield

$$
\phi_p \equiv \phi = \phi'' \qquad \text{on } \partial V_p' \tag{4.28}
$$

regardless of the values taken by the $A^{(\nu)}$'s. This means that even though we have not yet evaluated the $A^{(\nu)}$'s, we know that the last surface integral of Equation 4.26 will vanish upon substitution of Equation 4.27.

Second, we can observe that the mechanical surface force $\mathscr{F}_i$ is just a concentrated body force localized at ∂V_T. Thus, $\mathscr{F}_i$ and f_i can be combined symbolically into a single body force f_i' given by

$$
f_i' = f_i + \mathscr{F}_i \delta(V - \partial V_T) \tag{4.29}
$$

Finally, it is necessary to state two orthogonality relations which eigenmodes of different resonant frequencies can be shown to obey:

$$
\iiint_V \rho u_i^{(\nu)} u_i^{(\mu)} \, dV = \begin{cases} N^{(\mu,\nu)} & \omega_\nu = \omega_\mu \\ 0 & \text{otherwise} \end{cases} \tag{4.30}
$$

and

$$\iiint_V u_{i,j}^{(\nu)} c_{ijkl}^E u_{k,l}^{(\mu)}\, dV$$
$$+ \iiint_{V'} [\phi_{,m}^{(\nu)} e_{mkl} u_{k,l}^{(\mu)} + \phi_{,m}^{(\mu)} e_{mkl} u_{k,l}^{(\nu)} - \phi_{,m}^{(\nu)} \epsilon_{mn}^S \phi_{,n}^{(\mu)}]\, dV$$
$$= \begin{cases} N^{(\mu,\nu)}[\omega_\nu]^2 & \omega_\nu = \omega_\mu \\ 0 & \text{otherwise} \end{cases} \tag{4.31}$$

In Equation 4.30, the ω_ν is the resonant frequency, or eigenfrequency associated with mode ν and $N^{(\mu,\nu)}$ is the normalization integral. We shall assume hereafter that appropriate diagonalization procedures have been carried out in the event of frequency degeneracy and that the modes have been normalized in analogy with Equations 2.63 or 3.15 so as to make

$$N^{(\mu,\mu)} = 1 \tag{4.32}$$

If we now substitute Equation 4.27 into 4.26 and make use of Equations 4.28 to 4.32, we obtain

$$L = \frac{1}{2} \sum_\nu (A^{(\nu)})^2(\omega_\nu^2 - \omega^2) - \sum_\nu A^{(\nu)} \iiint_V u_i^{(\nu)} f'_i\, dV$$
$$+ \sum_\nu A^{(\nu)} \iiint_{V'} \phi''_{,m}[e_{mkl} u_{k,l}^{(\nu)} - \epsilon_{mn}^S \phi_{,n}^{(\nu)}]\, dV$$
$$- \frac{1}{2} \iiint_{V'} \phi''_{,m} \epsilon_{mn}^S \phi''_{,n}\, dV \tag{4.33}$$

At this point, we apply the stationary property of L to determine the $A^{(\nu)}$,

$$\frac{\partial L}{\partial A^{(\nu)}} = 0 \qquad \text{for all } \nu \tag{4.34}$$

Substitution of Equation 4.33 into 4.34, integration by parts of the second volume integral (*i.e.*, use of the divergence theorem), and use of Equations 4.13 and 4.17 as applied to mode ν then yield

$$A^{(\nu)} = \frac{1}{\omega_\nu^2 - \omega^2} \left\{ \iiint_V u_i^{(\nu)} f'_i\, dV \right.$$
$$\left. - \sum_{p=0}^{M} \iint_{\partial V_p'} \phi'' N_m [e_{mkl} u_{k\,l}^{(\nu)} - \epsilon_{mn}^S \phi_{,n}^{(\nu)}]\, dA \right\} \tag{4.35}$$

If we define $Q_p^{(\nu)}$ to be the free charge on electrode p associated with mode ν,

$$Q_p^{(\nu)} = -\iint\limits_{\partial V_p'} N_m[e_{mkl}u_{k,l}^{(\nu)} - \epsilon_{mn}^S \phi_{,n}^{(\nu)}]\, dA \tag{4.36}$$

we obtain from substitution of this definition and Equation 4.28 into Equation 4.35,

$$A^{(\nu)} = \frac{1}{\omega_\nu^2 - \omega^2}\left[\iiint\limits_V u_i^{(\nu)} f_i'\, dV + \sum_{p=1}^{M} \phi_p Q_p^{(\nu)}\right] \tag{4.37}$$

In this result, the summation begins with ϕ_1 rather than ϕ_0 by virtue of our convention that $\phi_0 = 0$.

Upon substitution of Equation 4.37 back into Equations 4.27, we obtain expressions for the displacement and potential response at every point of the system in Figure 4.1 expressed in terms of the natural modes $u_i^{(\nu)}$ and $\phi^{(\nu)}$ and the driving functions f_i' and ϕ_p. Before we can go on to obtain the admittance matrix formulas, it is necesssary also to express the current at electrode q,

$$I_q = -j\omega \iint\limits_{\partial V_q'} N_m D_m\, dA \tag{4.38}$$

in terms of these driving functions. This can be done by inserting Equations 4.2, 4.27, and 4.37 into Equation 4.38,

$$I_q = jw\left\{\sum_\nu \frac{Q_q^{(\nu)}}{\omega_\nu^2 - \omega^2}\left[\iiint\limits_V u_i^{(\nu)} f_i'\, dV + \sum_{p=1}^{M} \phi_p Q_p^{(\nu)}\right] + \iint\limits_{\partial V_q'} N_m \epsilon_{mn}^S \phi_{,n}''\, dA\right\} \tag{4.39}$$

Differentiation of I_q with respect to ϕ_p with f_i' held constant then yields the electrical portion of the desired admittance matrix,

$$Y_{qp} = \frac{\partial I_q}{\partial \phi_p} = j\omega \sum_\nu \frac{Q_q^{(\nu)} Q_p^{(\nu)}}{\omega_\nu^2 - \omega^2} + j\omega C_{qp} \tag{4.40*}$$

* Formulas very similar to this and Equation 4.60 have been obtained by Lloyd using a reciprocity integral theorem (see Reference 99).

In this relation,

$$C_{qp} = \iint\limits_{\partial V_q'} N_m \epsilon_{mn}^S \frac{\partial \phi''_{,n}}{\partial \phi_p} \, dA \tag{4.41}$$

may be identified as the clamped capacitance coupling between electrodes p and q.

Before developing the electromechanical and mechanical portions of the admittance matrix for the sample in Figure 4.1, we must define mechanical current and voltage analogues. In the present three-dimensional case, this is done by extending the two-dimensional procedure of Section 2.3. Let us assume, as in Equation 2.72, that the driving force density f_i' can be resolved into a linear superposition of distribution patterns $f_i^{(t)}$ multiplied by scalar weight factors ϕ_t'

$$f_i' = \sum_t f_i^{(t)} \phi_t' \tag{4.42}$$

These distribution patterns may be regarded as the mechanical terminals with the weight factors ϕ_t' as the associated voltage analogues. The analogue quantities associated with these mechanical terminals are to be indicated by the indices t (as in Equation 4.42) and w.

As an example of expansion Equation 4.42 consider a problem in which the external forces are only applied at discrete points. Then the $f_i^{(t)}$ may be defined as delta functions acting at those discrete points. Alternatively in the case of a distributed force, the $f_i^{(t)}$ may be defined as the particle displacement eigenmodes of the system $f_i^{(t)} = u_i^{(t)}$. In this second case, the summation of Equation 4.42 is understood to be infinite. Detailed discussions of these two (and other) choices for the $f_i^{(t)}$ have been given in Section 2.3 (for the two-dimensional case) and need not be repeated here.

Having defined the voltage analogues by Equation 4.42, we must exercise considerable caution in our choice of the current analogues I_w', or else our model will not retain the property of reciprocity inherent in the actual system. However, if we extend the two-dimensional definition, Equation 2.74, of I_w' to the present case, we may expect satisfactory results,

$$I_w' = j\omega \iiint\limits_V f_i^{(w)} u_i \, dV \tag{4.43}$$

We shall soon see that Equation 4.43 indeed leads to the desired reciprocity. Here, the $f_i^{(w)}$ are again the distribution patterns of Equation 4.42. For conciseness of notation, we shall now also define a "free mechanical charge of mode ν" analogue $Q_w'^{(\nu)}$ corresponding to $Q_p^{(\nu)}$ of Equation 4.36,

$$Q_w'^{(\nu)} = \iiint\limits_V f_i^{(w)} u_i^{(\nu)}\, dV \tag{4.44}$$

By means of the definitions, in Equations 4.41 to 4.44, we can express Equation 4.39 for I_q as

$$I_q = j\omega\left\{\sum_\nu \frac{Q_q^{(\nu)}}{\omega_\nu^2 - \omega^2}\left[\sum_t \phi_t' Q_t'^{(\nu)} + \sum_p \phi_p Q_p^{(\nu)}\right] + \sum_p C_{qp}\phi_p\right\} \tag{4.45}$$

Correspondingly, we can express the mechanical current analogue as

$$I_w' = j\omega\left\{\sum_\nu \frac{Q_w'^{(\nu)}}{\omega_\nu^2 - \omega^2}\left[\sum_t \phi_t' Q_t'^{(\nu)} + \sum_p \phi_p Q_p^{(\nu)}\right]\right\} \tag{4.46}$$

where Equations 4.27, 4.37, and 4.44 have been substituted into Equation 4.43.

The electromechanical portion of the admittance matrix is then obtained as the derivative of I_q with respect to ϕ_t' or as the derivative of I_w' with respect to ϕ_p,

$$\begin{aligned} Y_{qt} &= \frac{\partial I_q}{\partial \phi_t'} = j\omega \sum_\nu \frac{Q_q^{(\nu)} Q_t'^{(\nu)}}{\omega_\nu^2 - \omega^2} \\ Y_{wp} &= \frac{\partial I_w'}{\partial \phi_p} = j\omega \sum_\nu \frac{Q_p^{(\nu)} Q_w'^{(\nu)}}{\omega_\nu^2 - \omega^2} \end{aligned} \tag{4.47}$$

It should be remembered here that (t,w) index the mechanical variables and (p,q) index the electrical variables. Note that this result indeed possesses the required property of reciprocity.

Finally, the mechanical portion of the admittance matrix is given by

$$Y_{wt} = \frac{\partial I_w'}{\partial \phi_t'} = j\omega \sum_\nu \frac{Q_w'^{(\nu)} Q_t'^{(\nu)}}{\omega_\nu^2 - \omega^2} \tag{4.48}$$

We, of course, cannot represent the admittance matrix of a distributed parameter system such as we have in Figure 4.1 over all frequencies by a finite number of lumped parameter components. However, in the vicinity of the νth resonance, all the terms in the series of Equations 4.40, 4.47, and 4.48 except the νth may be negligible. If this is so, the admittance matrix of Equations 4.40, 4.47, and 4.48 for the sample in Figure 4.1 may be represented pictorially by the equivalent circuit of Figure 4.2. In this figure, because of the nature of the

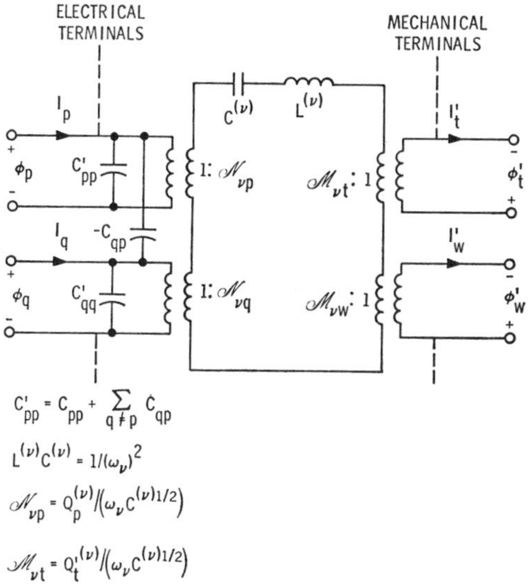

Figure 4.2. The equivalent circuit of the structure in Figure 4.1 near its νth short-circuit resonance.

transformer arrangement, one of the following is arbitrary: $C^{(\nu)}$, $L^{(\nu)}$, $\mathcal{N}_{\nu 1}$, $\mathcal{N}_{\nu 2}, \ldots, \mathcal{N}_{\nu M}$, $\mathcal{M}_{\nu 1}$, $\mathcal{M}_{\nu 2}, \ldots$. Consequently, one of these quantities may be assigned any value that is convenient. Once this has been done, the other element values are all specified by the formulas given in Figure 4.2.

This admittance matrix and equivalent circuit may be immediately observed to correspond with the Green's function results of Section 2.3, Equations 2.71, 2.73, 2.75, 2.76, and Figure 2.12. The only difference is that in the present three-dimensional case nonzero clamped transfer capacitances (C_{pq}, $p \neq q$) may be present.

The present results also may be given a rather interesting physical interpretation. In particular, we can regard Y_{qp} as the communication between terminals p and q, with the eigenmodes of the system acting as a medium for that communication. The individual terms in the series for Y_{qp} depend on the denominator resonance factor and on $Q_p^{(\nu)}$ and $Q_q^{(\nu)}$, which are measures of how well terminals p and q, respectively, are coupled to mode ν. This coupling is illustrated in Figure 4.2 by the proportionality between $Q_p^{(\nu)}$ and the turns ratio connecting electrode p to the resonant circuit of mode ν. Similar remarks apply to the electromechanical and mechanical admittances Y_{qt} and Y_{wt}.

The techniques described in this section are utilized later in this chapter to evaluate the equivalent circuit for a thick piezoelectric disk with concentric electrodes on the circular surfaces and for a rectangular parallelepiped with two opposite surfaces fully electroded.

4.2.3. The h-Matrix Formulation

Let us now consider the formulation of piezoelectric problems in which the elastic boundary conditions are unchanged from Equation 4.24, and in which we still have $\lambda = 0$ on $\partial V'_\lambda$. However, now let the total charge Q_r rather than the potential be specified on all the electrodes except the reference electrode $p = 0$. This situation corresponds to $M = 0$, $M' \neq 0$ in Equation 4.22. In this case, it is most convenient to evaluate the h-matrix of the problem, which consists of an impedance electrical submatrix, an admittance mechanical submatrix, and a mixed electromechanical submatrix.

Under the present boundary conditions, the Lagrangian of Equation 4.22 becomes

$$\begin{aligned} L = &\iiint_V \tfrac{1}{2}(u_{i,j}c^E_{ijkl}u_{k,l} - \rho\omega^2 u_i u_i)\, dV - \iiint_V u_i f_i \, dV \\ &+ \iiint_{V'} \tfrac{1}{2}(2\phi_{,m}e_{mkl}u_{k,l} - \phi_{,m}\epsilon^S_{mn}\phi_{,n})\, dV - \iint_{\partial V_T} \mathscr{F}_i u_i \, dA \\ &- \iint_{\partial V_0'} \phi N_m(e_{mkl}u_{k,l} - \epsilon^S_{mn}\phi_{,n})\, dA \\ &- \sum_{r=1}^{M'} \iint_{\partial V_r'} (\phi - \phi_r)N_m(e_{mkl}u_{k,l} - \epsilon^S_{mn}\phi_{,n})\, dA + \sum_{r=1}^{M'} \phi_r Q_r \end{aligned} \quad (4.49)$$

We again evaluate u_i and ϕ by expansion in the eigenmodes of the system $u_i^{(\nu,o)}$ and $\phi^{(\nu,o)}$, where these modes are now the homogeneous solutions of Equations 4.12 to 4.21 with f_i, ∂V_u, $\mathscr{F}_i$, λ, Q_r, and M all zero. These modes may be referred to as the open-circuit modes of the system, in contrast to the short-circuit modes used in the previous section. This distinction is reflected in the notation by the addition of an o in the superscripts. Thus, we assume

$$\begin{aligned} u_i &= \sum_\nu A^{(\nu)} u_i^{(\nu,o)} \\ \phi &= \sum_\nu A^{(\nu)} \phi^{(\nu,o)} + \phi'' \end{aligned} \quad (4.50)$$

and substitute this into Equation 4.49 in order to determine the $A^{(\nu)}$ coefficients.

This expansion will simplify Equation 4.49 considerably. First of all, ϕ'' and each $\phi^{(\nu,o)}$ are zero by definition on $\partial V_0'$. Consequently, upon substitution of Equations 4.50 the integral over $\partial V_0'$ in Equation 4.49 will vanish irrespective of the values taken by the $A^{(\nu)}$.

Moreover, let us define $\phi_r^{(\nu,o)}$ and ϕ_r'' to be the potential distributions $\phi^{(\nu,o)}$ and ϕ', respectively, evaluated on electrode r. Then, if we insert Equations 4.50 into the last surface integral of Equation 4.49, we obtain

$$-\sum_{r=1}^{M'}\left(\sum_{\nu} A^{(\nu)}\phi_r^{(\nu,o)} + \phi_r'' - \phi_r\right) \cdot \left\{\sum_{\mu} A^{(\mu)} \iint_{\partial V_r'} N_m(e_{mkl}u_{k,l}^{(\mu,o)} - \epsilon_{mn}^S \phi_{,n}^{(\mu,o)})\, dA - \iint_{\partial V_r'} N_m \epsilon_{mn}^S \phi_{,n}''\, dA\right\}$$

$$= \sum_{r=1}^{M'}\left(\sum_{\nu} A^{(\nu)}\phi_r^{(\nu,o)} + \phi_r'' - \phi_r\right) \iint_{\partial V_r'} N_m \epsilon_{mn}^S \phi_{,n}''\, dA \qquad (4.51)$$

The integrals of the terms associated with the $A^{(\mu)}$ are zero irrespective of the $A^{(\mu)}$, because they represent the free charge on the electrodes caused by the open-circuit modes; these quantities vanish by definition. This result can be further simplified by observing that the integral over $\partial V_r'$ in the remaining summation of Equation 4.51 is just the free charge on electrode r. Consequently, we find the last surface integral of Equation 4.49 becomes

$$-\sum_{r=1}^{M'} \iint_{\partial V_r'} (\phi - \phi_r) N_m(e_{mkl}u_{k,l} - \epsilon_{mn}^S \phi_{,n})\, dA$$

$$= \sum_{r=1}^{M'}\left(\sum_{\nu} A^{(\nu)}\phi_r^{(\nu,o)} + \phi_r'' - \phi_r\right) Q_r \qquad (4.52)$$

upon substitution of Equations 4.50, independently of the values ultimately assigned the $A^{(\nu)}$.

Thus, if u_i and ϕ are expanded as in Equations 4.50, the last three terms of Equation 4.49 reduce to $\Sigma_r(\Sigma_\nu A_r^{(\nu)}\phi^{(\nu,o)} + \phi_r'')Q_r$. In addition, orthogonality relations identical to Equations 4.30 and 4.31 hold for the $u_i^{(\nu,o)}$ and $\phi^{(\nu,o)}$. Hence, upon simplifying the last three terms of Equation 4.49 as indicated and then substituting Equations 4.29, 4.32, 4.50, and the orthogonality relations into the remaining terms of

Equation 4.49, we can derive the expression for L

$$
\begin{aligned}
L = \frac{1}{2}\sum_{\nu} (A^{(\nu)})^2(\omega_{\nu,o}^2 - \omega^2) - \sum_{\nu} A^{(\nu)} \iiint_V u_i^{(\nu,o)} f_i' \, dV \\
+ \sum_{\nu} A^{(\nu)} \iiint_{V'} \phi''_{,m}[e_{mkl} u_{k,l}^{(\nu,o)} - \epsilon_{mn}^S \phi_{,n}^{(\nu,o)}] \, dV \\
- \frac{1}{2} \iiint_{V'} \phi''_{,m} \epsilon_{mn}^S \phi''_{,n} \, dV + \sum_{r=1}^{M'} Q_r \left[\sum_{\nu} A^{(\nu)} \phi_r^{(\nu,o)} + \phi_r'' \right]
\end{aligned}
\tag{4.53}
$$

Here, the $\omega_{\nu,o}$ represent the open-circuit resonant frequencies.

If we now apply the stationary condition Equation 4.34, integrate the second volume integral by parts, and use Equations 4.13 and 4.17 as they pertain to open-circuit mode ν, we obtain the following expression for $A^{(\nu)}$:

$$
\begin{aligned}
A^{(\nu)} = \frac{1}{\omega_{\nu,o}^2 - \omega^2} \Bigg\{ \iiint_V u_i^{(\nu,o)} f_i' \, dV \\
- \sum_{r=1}^{M'} \phi_r'' \iint_{\partial V_r'} N_m [e_{mkl} u_{k,l}^{(\nu,o)} - \epsilon_{mn}^S \phi_{,n}^{(\nu,o)}] \, dA - \sum_{r=1}^{M'} Q_r \phi_r^{(\nu,o)} \Bigg\}
\end{aligned}
\tag{4.54}
$$

This equation may be condensed by observing, as in Equation 4.51, that the surface integral terms represent the identically zero free charge on the electrodes because of open-circuit mode ν. Thus, we have

$$
A^{(\nu)} = \frac{1}{\omega_{\nu,o}^2 - \omega^2} \left[\iiint_V u_i^{(\nu,o)} f_i' \, dV - \sum_{r=1}^{M'} Q_r \phi_r^{(\nu,o)} \right] \tag{4.55}
$$

Equations 4.50 and 4.55, like Equations 4.27 and 4.37, express the complete displacement and potential response in terms of the natural modes of the system and the driving functions.

Let us now evaluate the potential on electrode s in terms of the $A^{(\nu)}$ and the $\phi_r^{(\nu,o)}$. By virtue of Equations 4.50, we see that this potential ϕ_s is given by

$$
\phi_s = \sum_{\nu} A^{(\nu)} \phi_s^{(\nu,o)} + \phi_s'' \tag{4.56}
$$

The mechanical current and voltage analogues for this problem are again defined exactly as before in the short-circuit case (see Equations 4.42 and 4.43). However, the "free mechanical charge of mode ν"

analogue now becomes slightly modified from Equation 4.44,

$$Q_w'^{(\nu,o)} = \iiint_V f_i^{(w)} u_i^{(\nu,o)}\, dV \tag{4.57}$$

Thus, the potential at electrode s, ϕ_s can be expressed in terms of the natural modes and driving quantities by means of Equations 4.42, 4.55, 4.56, and 4.57 as

$$\phi_s = \sum_\nu \frac{\phi_s^{(\nu,o)}}{\omega_{\nu,o}^2 - \omega^2}\left[\sum_t \phi_t' Q_t'^{(\nu,o)} - \sum_{r=1}^{M'} \phi_r^{(\nu,o)} Q_r\right] + \phi_s'' \tag{4.58}$$

On the other hand, application of Equations 4.42, 4.43, 4.50, 4.55, and 4.57 gives the mechanical current analogue similarly in terms of the natural modes and driving terms as

$$I_w' = j\omega \sum_\nu \frac{Q_w'^{(\nu,o)}}{\omega_{\nu,o}^2 - \omega^2}\left[\sum_t \phi_t' Q_t'^{(\nu,o)} - \sum_{r=1}^{M'} \phi_r^{(\nu,o)} Q_r\right] \tag{4.59}$$

Differentiation of ϕ_s with respect to $I_r = j\omega Q_r$ then yields the electrical impedance submatrix,

$$\frac{\partial \phi_s}{\partial I_r} = Z_{sr} = \frac{1}{j\omega}\left[-\sum_\nu \frac{\phi_s^{(\nu,o)}\phi_r^{(\nu,o)}}{\omega_{\nu,o}^2 - \omega^2} + C_{sr}^{-1}\right] \tag{4.60}$$

In this result, C_{sr}^{-1}, which is given by

$$C_{sr}^{-1} = \frac{\partial \phi_s''}{\partial Q_r} \tag{4.61}$$

is the clamped transfer inverse capacitance. It is the matrix inverse of Equation 4.41.

The mechanical admittance submatrix may be obtained by differentiating I_w' with respect to ϕ_t':

$$Y_{wt} = \frac{\partial I_w'}{\partial V_t'} = j\omega \sum_\nu \frac{Q_w'^{(\nu,o)} Q_t'^{(\nu,o)}}{\omega_{\nu,o}^2 - \omega^2} \tag{4.62}$$

Finally, the electromechanical h submatrix is given by differentiating I_w' with respect to I_r, or by differentiating ϕ_s with respect to ϕ_t':

$$\begin{aligned} h_{wr} &= \frac{\partial I_w'}{\partial I_r} = -\sum_\nu \frac{Q_w'^{(\nu,o)}\phi_r^{(\nu,o)}}{\omega_{\nu,o}^2 - \omega^2} \\ h_{st} &= \frac{\partial \phi_s}{\partial \phi_t'} = \sum_\nu \frac{Q_t'^{(\nu,o)}\phi_s^{(\nu,o)}}{\omega_{\nu,o}^2 - \omega^2} \end{aligned} \tag{4.63}$$

It may be observed that this submatrix is antisymmetric rather than symmetric; this is characteristic of h-matrix representations of reciprocal systems.

It is possible to reduce the electrical or Z part of this response matrix to the result for multilayer plates that was derived by Green's function techniques in Section 2.4. The present response matrix is represented pictorially by an equivalent circuit which is considerably more complicated than that of Figure 4.2. This complexity greatly reduces its utility. For that reason, and because a previously published[68] equivalent circuit for the multilayer plate case does not differ materially from the equivalent circuit for the present case, we shall not further discuss equivalent circuits here.

4.2.4. Other Formulations

It is important to emphasize that the two foregoing formulations represent variations of the same answer to the same problem. The reason for selecting one over the other may be a matter of current I_p being easier to evaluate in a given case than voltage ϕ_p (so that the I_p make preferable independent variables) or *vice versa*. It may, on the other hand, be easier to evaluate the open-circuit modes occurring in the h-matrix than the short-circuit modes occurring in the admittance matrix (or *vice versa*).

Obviously, still other formulations may be derived that would prove advantageous in specific instances. For example, if part of the surface is clamped ($u_i = 0$ on ∂V_u), we have two options: We can use the admittance or h-matrix formulations that have been described here and then account for the effects of this clamping by open circuiting the appropriate mechanical analogue terminals (*i.e.*, by setting $I_t' = 0$). Alternatively, we can determine the natural modes of the system with this area clamped and then rederive the admittance or h-matrix using these new eigenmodes. The second approach would almost certainly involve less numerical computation.

The reader is, however, cautioned against using these techniques to derive formulations characterized by mechanical impedance rather than admittance submatrices. The problem in this case is that mechanical motion rather than force becomes the independent variable. Then, the relevant homogeneous boundary condition on the natural modes, instead of being zero stress on all of ∂V, is zero displacement on the ∂V_u portion of ∂V,

$$u_i^{(\nu)} = 0 \qquad \text{on } \partial V_u \tag{4.64}$$

These natural modes do not span the complete displacement function space, which is not restricted to $u_i = 0$ on ∂V_u. Consequently, this procedure will lead to severe convergence problems, such as Gibbs' phenomena, which can only be treated with extreme caution if at all.

4.3. Lagrangian Formulation for Determination of Electroelastic Normal Modes

4.3.1. Evaluation of the Short-Circuit Modes

The short circuit eigenmodes introduced in Equations 4.27 cannot in general be determined exactly. However, they may be approximated variationally using again the Lagrangian of Equation 4.22. In particular, substituting the relevant conditions of zero f_i, ∂V_u, $\mathscr{F}_i$, λ, ϕ_p, and M' reduces Equation 4.22 to

$$L = \iiint_V \tfrac{1}{2}(u_{i,j}c^E_{ijkl}u_{k,l} - \rho\omega^2 u_i u_i)\, dV + \iiint_{V'} \tfrac{1}{2}(2\phi_{,m}e_{mkl}u_{k,l} - \phi_{,m}\epsilon^S_{mn}\phi_{,n})\, dV - \sum_{p=0}^{M} \iint_{\partial V_p'} \phi N_m(e_{mkl}u_{k,l} - \epsilon^S_{mn}\phi_{,n})\, dA \tag{4.65}$$

We utilize this Lagrangian as in Chapter 3 (see Equation 3.8) by approximating the u_i and ϕ with a linear combination of trial functions,

$$u_i \doteq \sum_{\alpha=1}^{S_m} B^{(\alpha)} U_i^{(\alpha)} \qquad \phi \doteq \sum_{\beta=1}^{S_e} C^{(\beta)} \Phi^{(\beta)} \tag{4.66}$$

and substituting the result in Equation 4.65. Here, $B^{(\alpha)}$ and $C^{(\alpha)}$ are unknown coefficients to be determined. The trial functions $U_i^{(\alpha)}$ and $\Phi^{(\beta)}$ should all possess second derivatives. However, they need not satisfy any of the differential equations or boundary conditions, except that $U_i^{(\alpha)}$ should not be zero for all α and $\Phi^{(\beta)}$ should not be zero for all β at any point in V or on ∂V or $\partial V'$ where the exact solutions are not zero. Nevertheless, certain desirable results are achieved if the trial functions do satisfy some of the differential equations or boundary conditions, even though this is not required. For example, if the $\Phi^{(\beta)}$

satisfy $\phi = 0$ on $\partial V'_p$, the Lagrangian Equation 4.65 may be simplified by the omission of the last term. Also, as in the elastic case of Chapter 3, the convergence rate will be increased by picking trial functions that satisfy some or all of the boundary conditions. These statements will be reviewed in more detail later. Generally speaking, if the $U_i^{(\alpha)}$ and $\Phi^{(\beta)}$ are complete sets of functions, this method will converge to the exact eigenmodes of the problem as S_m and S_e go to infinity, provided the above nonzero restriction is not violated.*

If Equations 4.66 are substituted into Equation 4.65, we obtain the following approximation for L:

$$\begin{aligned} L \doteq \frac{1}{2}\sum_{\alpha=1}^{S_m}\sum_{\beta=1}^{S_m} B^{(\alpha)}B^{(\beta)}[P(\alpha,\beta) - \omega^2 H(\alpha,\beta)] \\ -\frac{1}{2}\sum_{\alpha=1}^{S_e}\sum_{\beta=1}^{S_e} C^{(\alpha)}C^{(\beta)}E(\alpha,\beta) + \sum_{\alpha=1}^{S_m}\sum_{\beta=1}^{S_e} B^{(\alpha)}C^{(\beta)}K(\alpha,\beta) \end{aligned} \tag{4.67}$$

The P (elastic), H (kinetic), E (electric), and K (piezoelectric) interaction matrices in this equation are given in Table 4.1. The $B^{(\alpha)}$ and $C^{(\beta)}$ are now evaluated by using the stationary property of the Lagrangian,

$$\begin{aligned} \frac{\partial L}{\partial B^{(\gamma)}} &= \sum_{\alpha=1}^{S_m} B^{(\alpha)}[P(\alpha,\gamma) - \omega^2 H(\alpha,\gamma)] + \sum_{\beta=1}^{S_e} C^{(\beta)}K(\gamma,\beta) = 0 \\ &\qquad \gamma = 1, 2, \ldots, S_m \\ \frac{\partial L}{\partial C^{(\gamma)}} &= -\sum_{\alpha=1}^{S_e} C^{(\alpha)}E(\alpha,\gamma) + \sum_{\alpha=1}^{S_m} B^{(\alpha)}K(\alpha,\gamma) = 0 \\ &\qquad \gamma = 1, 2, \ldots, S_e \end{aligned} \tag{4.68}$$

This operation corresponds to setting $\delta u_i = U_i^{(\gamma)}\delta B^{(\gamma)}$ and $\delta\phi = \Phi^{(\gamma)}\delta C^{(\gamma)}$ in Equation 4.23. The condition expresssed after Equation 4.22 that these δu_i and $\delta\phi$ need not satisfy the boundary conditions or differential equations gives us the previously mentioned liberty to let the $U_i^{(\gamma)}$ and $\Phi^{(\gamma)}$ violate the boundary conditions and differential equations. Like $\delta\phi$ and δu_i, the $\Phi^{(\gamma)}$ and $U_i^{(\gamma)}$ should, however, possess second derivatives in the domains of interest.

Equations 4.68 can be more concisely expressed in matrix notation,

$$\begin{aligned} [P - \omega^2 H][B] + [K][C] = 0 \\ -[E][C] + [K_t][B] = 0 \end{aligned} \tag{4.69}$$

* Sets of functions that are complete in the ordinary Fourier or normwise sense and which are not all zero at the same point are said to be pointwise complete.[74] Thus, {cos mx} forms a pointwise complete set on $0 \leq x \leq \pi$, but {sin mx} is only normwise complete. In most cases, our work will require trial function sets that are pointwise complete.

Table 4.1. The P, H, E, K, G, and J Interaction Matrices

$$P(\alpha,\beta) = \iiint_V U_{i,j}^{(\alpha)} c_{ijkl}^E U_{k,l}^{(\beta)}\, dV$$

$$H(\alpha,\beta) = \iiint_V U_i^{(\alpha)} \rho U_i^{(\beta)}\, dV$$

$$E(\alpha,\beta) = \iiint_{V'} \Phi_{,m}^{(\alpha)} \epsilon_{mn}^S \Phi_{,n}^{(\beta)}\, dV + \mathrm{SYM}[E'(\alpha,\beta)]$$

where $\mathrm{SYM}[E'(\alpha,\beta)]$ is the symmetric part of

$$E'(\alpha,\beta) = -2 \sum_{p=0}^{M} \iint_{\partial V_p'} \Phi^{(\alpha)} N_m \epsilon_{mn}^S \Phi_{,n}^{(\beta)}\, dA \text{ for the short-circuit modes}$$

$$E'(\alpha,\beta) = -2\left(\iint_{\partial V_0'} + \sum_{r=1}^{M'} \iint_{\partial V_r'}\right) \Phi^{(\alpha)} N_m \epsilon_{mn}^S \Phi_{,n}^{(\beta)}\, dA \text{ for the open-circuit modes}$$

$$K(\alpha,\beta) = \iiint_{V'} \Phi_{,m}^{(\beta)} e_{mkl} U_{k,l}^{(\alpha)}\, dV + K'(\alpha,\beta)$$

where

$$K'(\alpha,\beta) = -\sum_{p=0}^{M} \iint_{\partial V_p'} \Phi^{(\beta)} N_m e_{mkl} U_{k,l}^{(\alpha)}\, dA \text{ for the short-circuit modes}$$

$$K'(\alpha,\beta) = -\left(\iint_{\partial V_0'} + \sum_{r=1}^{M'} \iint_{\partial V_r'}\right) \Phi^{(\beta)} N_m e_{mkl} U_{k,l}^{(\alpha)}\, dA \text{ for the open-circuit modes}$$

$$G(r,\alpha) = \iint_{\partial V_r'} N_m e_{mkl} U_{k,l}^{(\alpha)}\, dA$$

$$J(r,\beta) = \iint_{\partial V_r'} N_m \epsilon_{mn}^S \Phi_{,n}^{(\beta)}\, dA$$

Here the subscript t designates a transposed matrix. Alternatively, Equations 4.69 may be represented as

$$\begin{aligned} &[C] = [E^{-1}K_t][B] \\ &[P + KE^{-1}K_t - \omega^2 H][B] = 0 \end{aligned} \tag{4.70}$$

It may be seen that this equation for the electroelastic short-circuit modes differs from Equation 3.13 for the purely elastic modes by the addition of the piezoelectric term $KE^{-1}K_t$.

Equations 4.70, like Equation 3.13, comprise a symmetric characteristic value problem, which possesses a solution only for discrete values of ω^2. These eigenvalues, of course, give the approximate short-circuit resonant frequencies of the system squared ω_ν^2.

Let us represent the eigenvectors associated with the ω_ν by $[B^{(\nu)}]$, and the individual components of $[B^{(\nu)}]$ by $B^{(\alpha,\nu)}$. These eigenvectors are determined only to a multiplicative constant by Equations 4.70. However, the approximate modal displacement patterns, which are specified by substitution of the $[B^{(\nu)}]$ into Equations 4.66,

$$u_i^{(\nu)} \doteq \sum_{\alpha=1}^{S_m} B^{(\alpha,\nu)} U_i^{(\alpha)} \tag{4.71}$$

are required to be normalized in accordance with Equations 4.30 and 4.32. Consequently, substitution of Equations 4.30 and 4.71 into Equation 4.32 leads to the following normalization condition on the $[B^{(\nu)}]$:

$$[B_t^{(\nu)}][H][B^{(\nu)}] = 1 \tag{4.72}$$

Here, the definition of H from Table 4.1 has also been applied.

Once the normalized $[B^{(\nu)}]$ of mode ν have been evaluated, the associated electric potential distribution may be found from Equations 4.66,

$$\phi^{(\nu)} \doteq \sum_{\beta=1}^{S_e} C^{(\beta,\nu)} \Phi^{(\beta)} \tag{4.73}$$

In this equation, the $C^{(\beta,\nu)}$ represent the components of $[C^{(\nu)}]$, which we determine from Equations 4.70 by

$$[C^{(\nu)}] = [E^{-1}K_t][B^{(\nu)}] \tag{4.74}$$

Two examples of short-circuit variational electroelastic modal calculations using the techniques of this section are presented later in this chapter.

4.3.2. *Evaluation of the Open-Circuit Modes*

It is also possible to obtain variational approximate solutions for the open-circuit eigenmodes that were introduced in Equation 4.50. The conditions of zero f_i, ∂V_u, $\mathscr{F}_i$, λ, Q_r, and M appropriate to this computation reduce the Lagrangian of Equation 4.22 to

$$
\begin{aligned}
L = & \iiint\limits_{V} \tfrac{1}{2}(u_{i,j}c^E_{ijkl}u_{k,l} - \rho\omega^2 u_i u_i)\, dV \\
& + \iiint\limits_{V'} \tfrac{1}{2}(2\phi_{,m}e_{mkl}u_{k,l} - \phi_{,m}\epsilon^S_{mn}\phi_{,n})\, dV \\
& - \iint\limits_{\partial V_0'} \phi N_m(e_{mkl}u_{k,l} - \epsilon^S_{mn}\phi_{,n})\, dA \\
& - \sum_{r=1}^{M'} \iint\limits_{\partial V_r'} (\phi - \phi_r) N_m(e_{mkl}u_{k,l} - \epsilon^S_{mn}\phi_{,n})\, dA
\end{aligned} \tag{4.75}
$$

We again expand u_i and ϕ in a linear combination of trial functions

$$
\begin{aligned}
u_i &\doteq \sum_{\alpha=1}^{S_m} B^{(\alpha,o)} U_i^{(\alpha)} \\
\phi &\doteq \sum_{\beta=1}^{S_e} C^{(\beta,o)} \Phi^{(\beta)}
\end{aligned} \tag{4.76}
$$

and substitute this into the Lagrangian. Again the $U_i^{(\alpha)}$ and $\Phi^{(\beta)}$ need not satisfy any of the boundary conditions or differential equations and may be quite arbitrary, except that they should not all be zero at any point where one does not anticipate the actual solution to be zero. Of course, the more of the boundary conditions or differential equations that are obeyed, the more rapid the convergence will be.

Performing the required substitution, we have the following approximation for L:

$$
\begin{aligned}
L \doteq & \frac{1}{2} \sum_{\alpha=1}^{S_m} \sum_{\beta=1}^{S_m} B^{(\alpha,o)} B^{(\beta,o)} [P(\alpha,\beta) - \omega^2 H(\alpha,\beta)] \\
& - \frac{1}{2} \sum_{\alpha=1}^{S_e} \sum_{\beta=1}^{S_e} C^{(\alpha,o)} C^{(\beta,o)} E(\alpha,\beta) + \sum_{\alpha=1}^{S_m} \sum_{\beta=1}^{S_e} B^{(\alpha,o)} C^{(\beta,o)} K(\alpha,\beta) \\
& + \sum_{r=1}^{M'} \sum_{\alpha=1}^{S_m} \phi_r B^{(\alpha,o)} G(r,\alpha) - \sum_{r=1}^{M'} \sum_{\beta=1}^{S_e} \phi_r C^{(\beta,o)} J(r,\beta)
\end{aligned} \tag{4.77}
$$

where the P, H, E, K, G, and J matrices are given in Table 4.1. Note that this Lagrangian changes from the corresponding short-circuit expression of Equation 4.67, primarily in the addition of the last two summations. The P, H, E, and K matrices do not differ in form from those of the previous formulations, except that the $U_i^{(\alpha)}$ and $\Phi^{(\beta)}$ trial functions most efficient here (Equations 4.76) would not be the same as those of the short-circuit case (Equations 4.66).

Let us represent the $B^{(\alpha,o)}$ and $C^{(\beta,o)}$ arrays by $[B^{(o)}]$ and $[C^{(o)}]$, and let us additionally designate the ϕ_r array by $[\phi]$. Use of the stationary property of the Lagrangian in this case yields three sets of equations,

$$\frac{\partial L}{\partial B^{(\gamma,o)}} = \frac{\partial L}{\partial C^{(\gamma,o)}} = \frac{\partial L}{\partial \phi_s} = 0 \tag{4.78}$$

Substitution of Equation 4.77 into Equation 4.78 then gives three matrix equations,

$$\begin{aligned} [P - \omega^2 H][B^{(o)}] + [K][C^{(o)}] + [G_t][\phi] &= 0 \\ -[E][C^{(o)}] + [K_t][B^{(o)}] - [J_t][\phi] &= 0 \\ [G][B^{(o)}] - [J][C^{(o)}] &= 0 \end{aligned} \tag{4.79}$$

If the second of these is solved for $[C]$, we obtain

$$[C^{(o)}] = [E^{-1}K_t][B^{(o)}] - [E^{-1}J_t][\phi] \tag{4.80}$$

and

$$\begin{aligned} [P + KE^{-1}K_t - \omega^2 H][B^{(o)}] + [G_t - KE^{-1}J_t][\phi] &= 0 \\ [G - JE^{-1}K_t][B^{(o)}] + [JE^{-1}J_t][\phi] &= 0 \end{aligned} \tag{4.81}$$

Finally, solving the last of these for $[\phi]$ we have

$$[\phi] = -[JE^{-1}J_t]^{-1}[G - JE^{-1}K_t][B^{(o)}] \tag{4.82}$$

and

$$\begin{aligned} [P + KE^{-1}K_t - (G_t - KE^{-1}J_t)(JE^{-1}J_t)^{-1} \\ (G - JE^{-1}K_t) - \omega^2 H][B^{(o)}] = 0 \end{aligned} \tag{4.83}$$

Equation 4.83 comprises a symmetric characteristic value problem with nontrivial solutions only for $\omega = \omega_{\nu,o}$, the approximate open-circuit resonant frequencies. The associated eigenvectors $[B^{(\nu,o)}]$, when normalized in accordance with Equation 4.72, yield the approximate open-circuit modal displacements,

$$u_i^{(\nu,o)} \doteq \sum_{\alpha=1}^{S_m} B^{(\alpha,\nu,o)} U_i^{(\alpha)} \tag{4.84}$$

The νth eigenvector also yields the $[\phi]$ of mode ν, $[\phi^{(\nu)}]$, when substituted into Equation 4.82,

$$[\phi^{(\nu)}] = -[JE^{-1}J_t]^{-1}[G - JE^{-1}K_t][B^{(\nu,o)}] \tag{4.85}$$

The elements of this array are approximately the open-circuit electrode voltages of mode ν, which appear in the h-parameter matrix summations of Equations 4.60 and 4.63,

$$\phi_r^{(\nu,o)} = \phi_r^{(\nu)} \tag{4.86}$$

Finally, substitution of $[B^{(\nu,o)}]$ and $[\phi^{(\nu)}]$ into Equation 4.80 determines the $[C^{(\nu,o)}]$ array of open-circuit mode ν, which in turn gives the potential distribution pattern $\phi^{(\nu,o)}$ by substitution into Equations 4.76:

$$\phi^{(\nu,o)} \doteq \sum_{\beta=1}^{S_e} C^{(\beta,\nu,o)} \phi^{(\beta)} \tag{4.87}$$

4.4. Application to Rectangular Piezoelectric Parallelepipeds

4.4.1. Theory

Let us now illustrate these variational techniques by applying them to the rectangular ferroelectric ceramic parallelepiped shown with dimensions and coordinate system in Figure 4.3. We will assume the top and bottom surfaces are completely covered by thin electrodes and that the sample is everywhere mechanically free.

Earlier literature on this type of problem is somewhat scanty. About the only work in print seems to be that of Ekstein and Schiffmann, who evaluated very approximately the natural modes of isotropic (nonpiezoelectric) cubes and nearly cubic parallelepipeds.[100,101]

The first step in analyzing a piezoelectric structure such as this is to determine the natural modes. Moreover, the short-circuit modes which lead to the admittance formula described in Section 4.2.2 are more

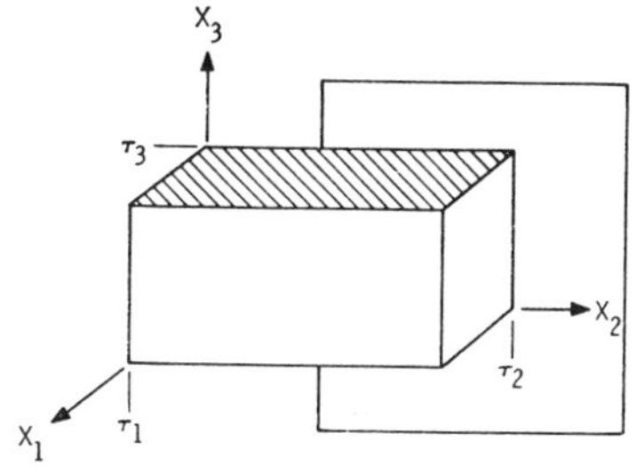

Figure 4.3. A rectangular ferroelectric ceramic parallelepiped. Note the external short circuit. Polarization is in the x_3 direction.

useful in this case than the alternative open-circuit modes. Consequently, we will begin by considering the two electroded surfaces in Figure 4.3 to be externally shorted together.

Under these conditions, the short-circuit Lagrangian of Equation 4.65 applies. In the present geometry, this Lagrangian takes the form

$$L = \iiint_V \tfrac{1}{2}(u_{i,j}c^E_{ijkl}u_{k,l} + 2\phi_{,m}e_{mkl}u_{k,l} - \phi_{,m}\epsilon^S_{mn}\phi_{,n} - \rho\omega^2 u_i u_i)\,dV$$
$$- \iint_A [\phi(e_{3kl}u_{k,l} - \phi_{,n}\epsilon^S_{3n})]^{x_3=\tau_3}_{x_3=0}\,dx_1\,dx_2 \qquad (4.88)$$

where A is the sample cross section in the x_1x_2-plane and electrode volume $V - V'$ is neglected.

This Lagrangian is utilized as outlined in the preceding section by expanding u_i and ϕ (see Equations 4.66) to obtain an approximate formula for L as in Equation 4.67. Carrying out manipulations 4.68 and 4.69 then, we have the result that the νth short circuit mode of the parallelepiped is described approximately by

$$\begin{aligned} u_i^{(\nu)} &\doteq \sum_{\alpha=1}^{S_m} B^{(\alpha,\nu)}U_i^{(\alpha)} \\ \phi^{(\nu)} &\doteq \sum_{\beta=1}^{S_e} C^{(\beta,\nu)}\Phi^{(\beta)} \end{aligned} \qquad (4.89)$$

Here the arrays $[B^{(\nu)}]$ and $[C^{(\nu)}]$ are the νth solutions of the characteristic value problem, Equations 4.70,

$$\begin{aligned} &[C^{(\nu)}] = [E^{-1}K_t][B^{(\nu)}] \\ &[P + KE^{-1}K_t - \omega_\nu^2 H][B^{(\nu)}] = 0 \end{aligned} \qquad (4.90)$$

In Equations 4.89 and 4.90, the $[B^{(\nu)}]$ is understood to be normalized according to Equation 4.72.

Numerical studies carried out by the authors of this monograph employed six types of trial functions in Equations 4.89 to expand the particle displacement u_i and two types to expand the electric potential ϕ. The general expressions for these eight trial function types are presented in Table 4.2.

If the external short circuit in Figure 4.3 is removed and a sinusoidal voltage applied in its place, one can expect to excite only modes in the parallelepiped having dilation type symmetry:

$$
\begin{array}{ll}
u_1^{(\nu)} & \text{odd in } x_1\text{, even in } x_2 \text{ and } x_3 \text{ with respect to the center of } V \\
u_2^{(\nu)} & \text{odd in } x_2\text{, even in } x_1 \text{ and } x_3 \text{ with respect to the center of } V \\
u_3^{(\nu)} & \text{odd in } x_3\text{, even in } x_1 \text{ and } x_2 \text{ with respect to the center of } V \\
\phi^{(\nu)} & \text{odd in } x_3\text{, even in } x_1 \text{ and } x_2 \text{ with respect to the center of } V
\end{array}
\tag{4.91}
$$

This is true because shear is not coupled to axial electric field in a ferroelectric ceramic. The trial functions given in Table 4.2 have all been selected so as to possess this dilation type of symmetry. In addition, we may observe that all the electric potential trial functions satisfy the $\phi = 0$ boundary conditions on the electrodes. Although this is not, as we pointed out after Equations 4.68, a necessary condition on the $\Phi^{(\beta)}$, it will accelerate convergence. Also, it will simplify the K and E matrices given in Table 4.1 by reducing to zero the surface integral terms.

As with the set of trial functions of Equations 3.37 and 3.38 used in analyzing rectangular plates, the present set is pointwise complete and, in the normwise sense, over-complete.* In fact, we should be able to obtain convergent modal short-circuit solutions to the parallelepiped with trial functions of the first, third, fifth, and seventh type in Table 4.2 omitted altogether: What would then remain still forms a pointwise complete set. However, as in the rectangular plate case, we have found that without the additional terms convergence is too slow to give meaningful answers even with the most modern of high-speed computers. As explained after Equations 3.37, the problem is that without the over-complete contribution it takes an unreasonable number of terms to approximate closely the mechanically free boundary condition ($\mathscr{F}_i = 0$ in Equation 4.15).

The analysis now proceeds by substituting the trial functions of Table 4.2 into the matrix formulas of Table 4.1 (with the surface integrals in K and E omitted). We also apply the symmetry requirements of a ferroelectric ceramic (C_{6v} in Table 1.1) to the coefficient

* Sets of functions that are complete in the ordinary Fourier or normwise sense and which are not all zero at the same point are said to be pointwise complete.[74] Thus, $\{\cos mx\}$ forms a pointwise complete set on $0 \leq x \leq \pi$, but $\{\sin mx\}$ is only normwise complete.

Table 4.2. General Expressions for the Trial Function Types

Trial Function	Range of α or β	Comments
$U_i^{(\alpha)} = \delta_{i1} \cos \frac{m_{\alpha 1}\pi x_1}{\tau_1} \sin \frac{n_{s\alpha 1}\pi x_2}{\tau_2} \sin \frac{p_{s\alpha 1}\pi x_3}{\tau_3}$ $m_{\alpha 1}, n_{s\alpha 1}, p_{s\alpha 1}$ all odd	$\alpha = 1, 2, \ldots, n_{s1T}$	δ_{i1} is Kronecker delta; thus only x_1 component is nonzero
$U_i^{(\alpha)} = \delta_{i1} \cos \frac{m_{\alpha 1}\pi x_1}{\tau_1} \cos \frac{n_{c\alpha 1}\pi x_2}{\tau_2} \cos \frac{p_{c\alpha 1}\pi x_3}{\tau_3}$ $m_{\alpha 1}$ odd, $n_{c\alpha 1}$ and $p_{c\alpha 1}$ even	$\alpha = n_{s1T} + (1, 2, \ldots, n_{c1T})$	only x_1 component nonzero
$U_i^{(\alpha)} = \delta_{i2} \sin \frac{m_{s\alpha 2}\pi x_1}{\tau_1} \cos \frac{n_{\alpha 2}\pi x_2}{\tau_2} \sin \frac{p_{s\alpha 2}\pi x_3}{\tau_3}$ $m_{s\alpha 2}, n_{\alpha 2}, p_{s\alpha 2}$ all odd	$\alpha = n_{s1T} + n_{c1T} + (1, 2, \ldots, m_{s2T})$ $= m_{1T} + (1, 2, \ldots, m_{s2T})$	only x_2 component nonzero
$U_i^{(\alpha)} = \delta_{i2} \cos \frac{m_{c\alpha 2}\pi x_1}{\tau_1} \cos \frac{n_{\alpha 2}\pi x_2}{\tau_2} \cos \frac{p_{c\alpha 2}\pi x_3}{\tau_3}$ $n_{\alpha 2}$ odd, $m_{c\alpha 2}$ and $p_{c\alpha 2}$ even	$\alpha = m_{1T} + m_{s2T} + (1, 2, \ldots, m_{c2T})$	only x_2 component nonzero

$U_i^{(\alpha)} = \delta_{i3} \sin \dfrac{m_{s\alpha 3}\pi x_1}{\tau_1} \sin \dfrac{n_{s\alpha 3}\pi x_2}{\tau_2} \cos \dfrac{p_{\alpha 3}\pi x_3}{\tau_3}$	$\alpha = m_{1T} + m_{s2T} + m_{c2T} + (1, 2, \ldots, m_{s3T})$ $= m_{1T} + n_{2T} + (1, 2, \ldots, m_{s3T})$	only x_3 component nonzero
$m_{s\alpha 3}, n_{s\alpha 3}, p_{\alpha 3}$ all odd		
$U_i^{(\alpha)} = \delta_{i3} \cos \dfrac{m_{c\alpha 3}\pi x_1}{\tau_1} \cos \dfrac{n_{c\alpha 3}\pi x_2}{\tau_2} \cos \dfrac{p_{\alpha 3}\pi x_3}{\tau_3}$	$\alpha = m_{1T} + n_{2T} + m_{s3T} + (1, 2, \ldots, m_{c3T})$	only x_3 component nonzero
$p_{\alpha 3}$ odd, $m_{c\alpha 3}$ and $n_{c\alpha 3}$ even	$(m_{1T} + n_{2T} + m_{s3T} + m_{c3T} = S_m)$	
$\Phi^{(\beta)} = \sin \dfrac{m_{s\beta 4}\pi x_1}{\tau_1} \sin \dfrac{n_{s\beta 4}\pi x_2}{\tau_2} \sin \dfrac{p_{\beta 4}\pi x_3}{\tau_3}$	$\beta = 1, 2, \ldots, m_{s4T}$	$\Phi^{(\beta)} = 0$ at $x_3 = 0, \tau_3$
$m_{s\beta 4}$ and $n_{s\beta 4}$ odd, $p_{\beta 4}$ even		
$\Phi^{(\beta)} = \cos \dfrac{m_{c\beta 4}\pi x_1}{\tau_1} \cos \dfrac{n_{c\beta 4}\pi x_2}{\tau_2} \sin \dfrac{p_{\beta 4}\pi x_3}{\tau_3}$	$\beta = m_{s4T} + (1, 2, \ldots, m_{c4T})$	$\Phi^{(\beta)} = 0$ at $x_3 = 0, \tau_3$
$m_{c\beta 4}, n_{c\beta 4}, p_{\beta 4}$ all even	$(m_{s4T} + m_{c4T} = S_e)$	

tensors appearing in the integrands of Table 4.1. These operations lead to recursion formulas for the elements of 36 submatrices of P and H, 12 submatrices of K, and 4 submatrices of E, where the various submatrices represent all possible interactions of the eight types of trial functions. The resulting recursion formulas are tabulated elsewhere.[102]

Table 4.3. Values of m's, n's, and p's Characterizing the Present Trial Functions*

A. Mechanical Trial Functions	
$(m_{\alpha 1}, n_{s\alpha 1}, p_{s\alpha 1})$ $= (n_{\alpha 2}, p_{s\alpha 2}, m_{s\alpha 2})$ $= (p_{\alpha 3}, m_{s\alpha 3}, n_{s\alpha 3})$	$(m_{\alpha 1}, n_{c\alpha 1}, p_{c\alpha 1})$ $= (n_{\alpha 2}, p_{c\alpha 2}, m_{c\alpha 2})$ $= (p_{\alpha 3}, m_{c\alpha 3}, n_{c\alpha 3})$
(1,1,1), (3,1,1), (5,1,1), (7,1,1), (9,1,1), (11,1,1), (13,1,1), (15,1,1)	(1,0,0), (1,0,2), (1,0,4), (1,0,6), (1,0,8), (1,2,0), (1,2,2), (1,2,4), (1,4,0), (1,4,2), (1,6,0), (1,8,0), (3,0,0), (3,0,2), (3,0,4), (3,0,6), (3,2,0), (3,2,2), (3,4,0), (3,6,0), (5,0,0), (5,0,2), (5,2,0), (7,0,0), (9,0,0)
B. Electrical Trial Functions	
$(p_{\beta 4}, m_{s\beta 4}, n_{s\beta 4})$	$(p_{\beta 4}, m_{c\beta 4}, n_{c\beta 4})$
(2,1,1), (4,1,1), (6,1,1), (8,1,1), (10,1,1), (12,1,1), (14,1,1), (16,1,1)	(2,0,0), (2,0,2), (2,0,4), (2,0,6), (2,0,8), (2,2,0), (2,2,2), (2,2,4), (2,4,0), (2,4,2), (2,6,0), (2,8,0), (4,0,0), (4,0,2), (4,0,4), (4,0,6), (4,2,0), (4,2,2), (4,4,0), (4,6,0), (6,0,0), (6,0,2), (6,2,0), (8,0,0), (10,0,0)

* The trial functions used in this work result from substituting the integer triplets of this table into the general expressions for the eight trial function types given in Table 4.2.

It has been found that about 3 percent accurate values for $\omega_\nu (\nu < 30)$ may be obtained with 33 trial functions in the expansion of each component of u_i and 33 trial functions in the expansion for ϕ. The m's, n's, and p's which characterize these trial functions are given in Table 4.3. All subsequently reported numerical results are based on this set of trial functions.

In the interest of limiting the scope of this example by reasonable bounds, let us restrict our attention to samples of square cross section ($\tau_1 = \tau_2$ in Figure 4.3), although all the theory presented so far is not itself so restricted in this way. For such samples, as in the case of thin square plates, it is found that all dilation modes are either in-phase or out-of-phase. The in-phase modes are characterized by

$$\begin{aligned} u_1^{(\nu)}(x_1,x_2,x_3) &= u_2^{(\nu)}(x_2,x_1,x_3) \\ u_3^{(\nu)}(x_1,x_2,x_3) &= u_3^{(\nu)}(x_2,x_1,x_3) \end{aligned} \tag{4.92}$$

and the out-of-phase modes by

$$\begin{aligned} u_1^{(\nu)}(x_1,x_2,x_3) &= -u_2^{(\nu)}(x_2,x_1,x_3) \\ u_3^{(\nu)}(x_1,x_2,x_3) &= -u_3^{(\nu)}(x_2,x_1,x_3) \end{aligned} \tag{4.93}$$

Let us define the normalized or dimensionless resonant frequency of mode ν as

$$\Omega_\nu^2 = \frac{\omega_\nu^2 \tau_1^2 \rho}{\pi^2 c_{3333}^E} \tag{4.94}$$

Figure 4.4 then shows the normalized spectrum for the short-circuit dilation modes of a lead-zirconate-titanate ferroelectric ceramic (PZT-5) square cylinder as the thickness-to-width ratio is varied from zero to one. Electroelastic coefficients utilized in this computation are from Table 1.3. It is interesting to note that in-phase modes never cross one another nor do out-of-phase modes. This situation illustrates the group theory conclusion that only the resonant frequencies of modes belonging to different irreducible representations can cross.[101,103]

To be a bit more specific, the symmetry group of a ferroelectric ceramic square cylinder is C_{4v}. Under all the operations of this group, the in-phase modes transform as the quantity $x_1^2 + x_2^2$; they are completely invariant.[104,105] On the other hand, the out-of-phase modes transform as $x_1^2 - x_2^2$; they are invariant under operations E, C_4^2, and σ_v, but they are multiplied by -1 under operations C_4, C_4^3, and σ_d. Consequently, the in-phase modes belong to the irreducible representation A_1, and the out-of-phase modes belong to a different representation B_1.[104,105]

An isotropic cube may be regarded as a special case of a ferroelectric ceramic square cylinder. In this special case, a considerably higher degree of symmetry is found, that of group O_h. Figure 4.5 shows the normalized resonant frequencies of the first 50 dilation modes as a function of Poisson's ratio for an isotropic cube. Here, all modes fall into one of four subfamilies, rather than just two as above.

For example, some modes are found to occur in degenerate pairs or doublets. One member of these doublets is always an in-phase mode, as represented by Equations 4.92, and one is always out-of-phase, as represented by Equations 4.93. These doublets belong to the E_g

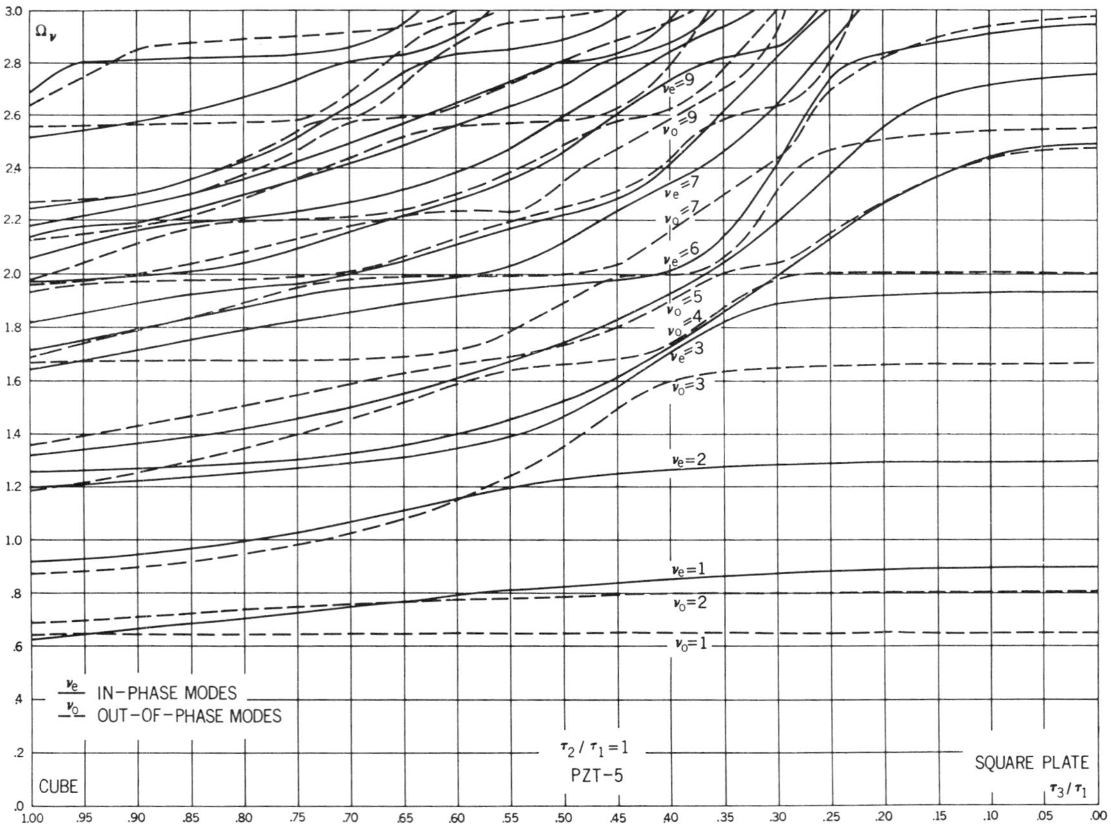

Figure 4.4. Normalized spectrum of the dilation modes for a lead-zirconate-titanate ceramic (PZT-5) square cylinder. The thickness-to-width ratio is varied from unity (cube case) to zero (thin square plate case). In-phase modes are indexed by ν_e and out-of-phase modes by ν_0.

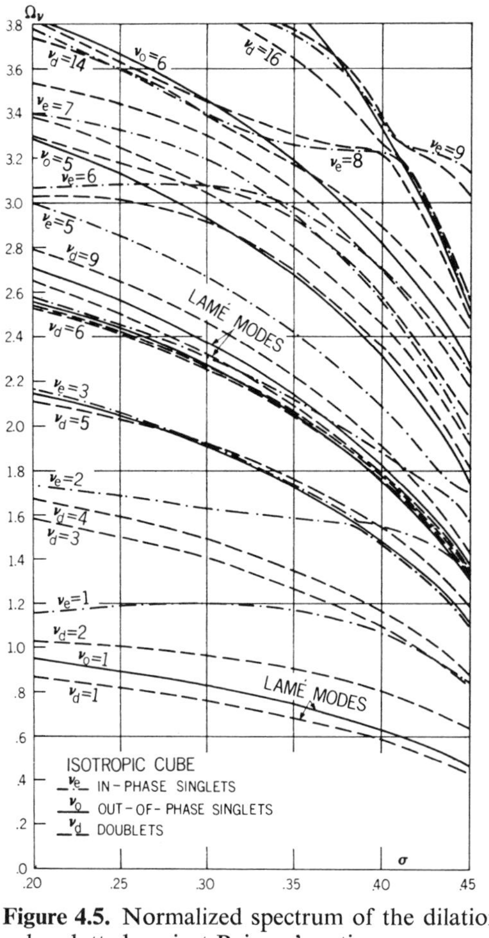

Figure 4.5. Normalized spectrum of the dilation modes of an isotropic cube plotted against Poisson's ratio.

irreducible representation of group O_h.[105,106] They are represented by dashed lines in Figure 4.5 and indexed with ν_d.

Additional cube modes occur singly and are invariant to all operations in O_h,

$$u_1^{(\nu)}(x_1,x_2,x_3) = u_1^{(\nu)}(x_1,x_3,x_2) = u_2^{(\nu)}(x_2,x_1,x_3) = u_2^{(\nu)}(x_2,x_3,x_1) = u_3^{(\nu)}(x_3,x_1,x_2) = u_3^{(\nu)}(x_3,x_2,x_1) \tag{4.95}$$

These modes constitute a special case of the in-phase subfamily described by Equations 4.92. They are indexed by ν_e in Figure 4.5, and belong to representation A_{1g}.[105,106]

Other singlet modes are invariant to operations E, C_3, C_4^2, i, S_6, and σ_h, but are multiplied by -1 under C_2', C_4, C_4^3, S_4, and σ_d,

$$\begin{aligned} u_1^{(v)}(x_1,x_2,x_3) &= -u_1^{(v)}(x_1,x_3,x_2) = u_2^{(v)}(x_2,x_3,x_1) = -u_2^{(v)}(x_2,x_1,x_3) \\ &= u_3^{(v)}(x_3,x_1,x_2) = -u_3^{(v)}(x_3,x_2,x_1) \end{aligned} \tag{4.96}$$

These modes form a special case of the out-of-phase subfamily (Equations 4.93) and are indexed by ν_c in Figure 4.5. They belong to representation A_{2g}.[105,106]

It may be observed in Figure 4.5 that again the frequencies of modes belonging to the same irreducible representation do not cross but that those belonging to different representations do cross.

In Figures 4.6 to 4.10 are shown level curves of the x_3-displacement at $x_3 = 0$ for the first out-of-phase mode and first four in-phase modes of (1) an isotropic cube with $\sigma = 0.30$, (2) a PZT-5 cube, (3) a PZT-5 parallelepiped with $\frac{3}{4}\tau_1 = \frac{3}{4}\tau_2 = \tau_3$, and (4) a PZT-5 parallelepiped with $\frac{1}{2}\tau_1 = \frac{1}{2}\tau_2 = \tau_3$. The distinction between in-phase and out-of-phase modes is readily apparent here; the main diagonals are nodal

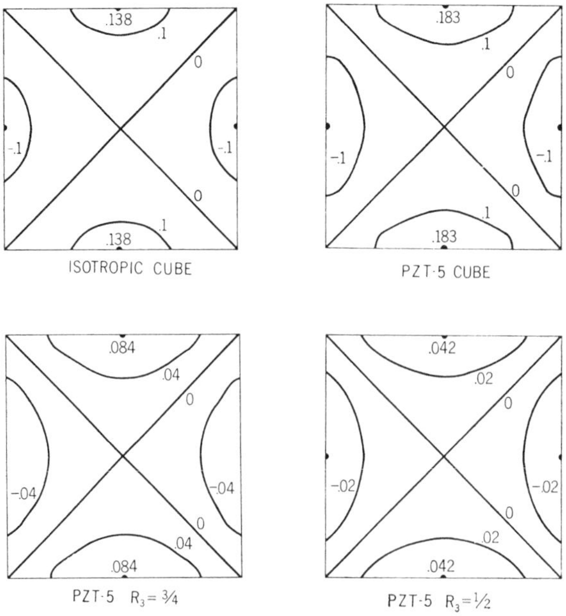

Figure 4.6. Level curves of $\psi_3^{(v)}$ at $x_3 = 0$ for the first out-of-phase dilation mode; $\nu_d = 1$ for isotropic samples and $\nu_0 = 1$ for PZT-5 samples, $R_3 = \tau_3/\tau_1$.

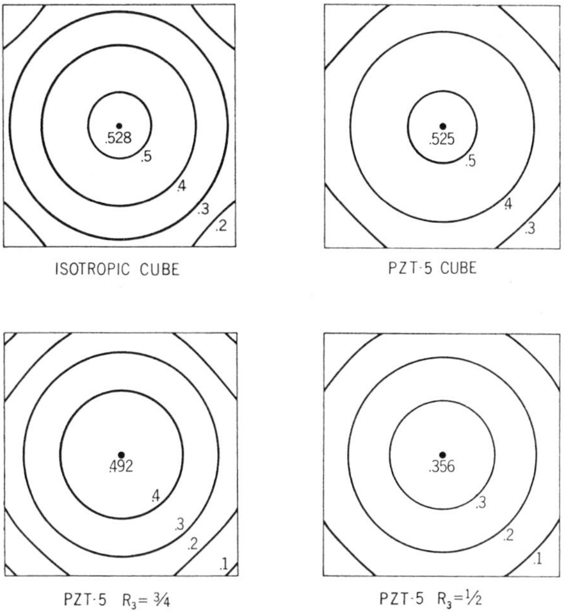

Figure 4.7. Level curves of $\psi_3^{(\nu)}$ at $x_3 = 0$ for the first in-phase dilation mode; $\nu_d = 1$ for the isotropic sample and $\nu_e = 1$ for PZT-5 samples, $R_3 = \tau_3/\tau_1$.

lines for the out-of-phase mode. These figures pertain to modes normalized according to Equations 4.72 and 4.90, and then renormalized into dimensionless form by the rule

$$\psi_i^{(\nu)} = \tau_1 \pi^{-1} (\rho \tau_3)^{1/2} u_i^{(\nu)} \qquad (4.97)$$

(Compare this with Equation 3.55).

4.4.2. *Determination of the Dynamic Capacitances*

When the external short circuit in Figure 4.3 is removed, the sample may be represented electrically over all frequencies by the equivalent circuit of Figure 4.11. This circuit is a pictorial representation of Equation 4.40,

$$Y = j\omega C_S + j\omega \sum_{\nu} \frac{(Q^{(\nu)})^2}{\omega_\nu^2 - \omega^2} \qquad (4.98)^*$$

* Subscripts on Y and $Q^{(\nu)}$ are dropped in this section, as the present sample will possess only one ungrounded electrode. Let us assume the bottom or $x_3 = 0$ electrode in Figure 4.3 is the grounded electrode O. Then the top electrode is electrode 1; Y means Y_{11}; and $Q^{(\nu)}$ means $Q_1^{(\nu)}$. This convention will apply throughout this section.

as it applies to the piezoelectric parallelepiped. Note that Figure 4.11 does not differ qualitatively from Figure 3.7, the corresponding result for a thin rectangular plate.

In Figure 4.11 or Equation 4.98, the C_S is the clamped capacitance; *i.e.*, the capacitance with all mechanical motion prevented,

$$C_S = \frac{\epsilon_{33}^S \tau_1 \tau_2}{\tau_3} \tag{4.99}*$$

Furthermore, the infinite series of LC resonators in Figure 4.11 describe the resonances of the sample,

$$\omega_\nu^2 = \frac{1}{L_\nu C_\nu} \tag{4.100}$$

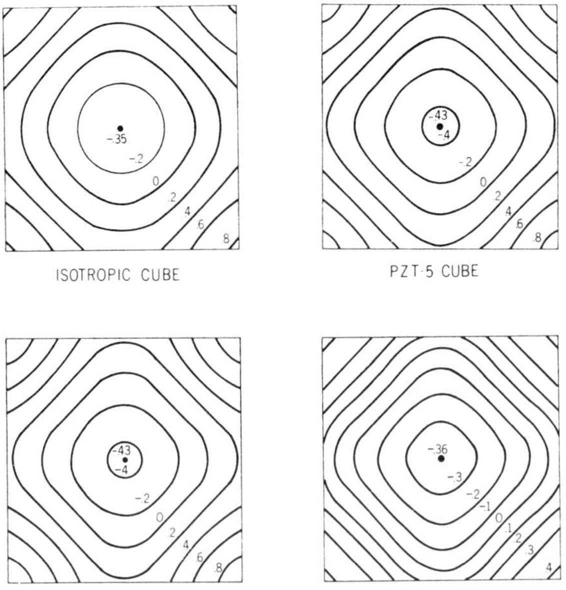

Figure 4.8. Level curves of $\psi_3^{(\nu)}$ at $x_3 = 0$ for the second in-phase dilation mode; $\nu_d = 2$ for the isotropic sample and $\nu_e = 2$ for the PZT-5 samples, $R_3 = \tau_3/\tau_1$.

* C_S of Equation 4.99 differs from C_0 of Equation 3.49 in that Equation 3.49 is evaluated with the sample clamped only in the x_1x_2-plane. The present C_S is evaluated with motion prevented in all three directions.

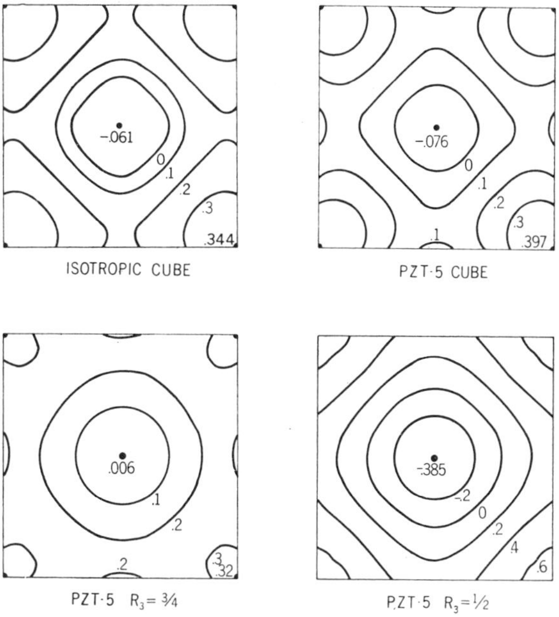

Figure 4.9. Level curves of $\psi_3^{(\nu)}$ at $x_3 = 0$ for the third in-phase dilation mode; $\nu_e = 1$ for the isotropic sample and $\nu_e = 3$ for the PZT-5 samples, $R_3 = \tau_3/\tau_1$.

The quantity C_ν is often referred to as the dynamic or motional capacitance of mode ν.

At near zero frequency, the sample in Figure 4.3 is representable as a single capacitor of value specified by the mechanically free dielectric constant (Equation 1.8 or 1.12),

$$C_T = \frac{\epsilon_{33}^T \tau_1 \tau_2}{\tau_3} \tag{4.101}$$

Consequently, since all the L_ν look like short circuits near zero frequency, all C_ν of Figure 4.11 must sum up to the difference between C_T and C_S,

$$C_T - C_S = \sum_\nu C_\nu = (\epsilon_{33}^T - \epsilon_{33}^S)\tau_1\tau_2/\tau_3 \tag{4.102}$$

(A bit of algebra indicates that this difference may also be expressed as

$$C_T - C_S = [k^{(5)}]^2 \epsilon_{33}^T \tau_1 \tau_2/\tau_3 \tag{4.103}$$

where this $k^{(5)}$ is the eigen coupling factor of Equations 1.90. Liberal use of Equations 1.24 and 1.26 is required to establish this result.)

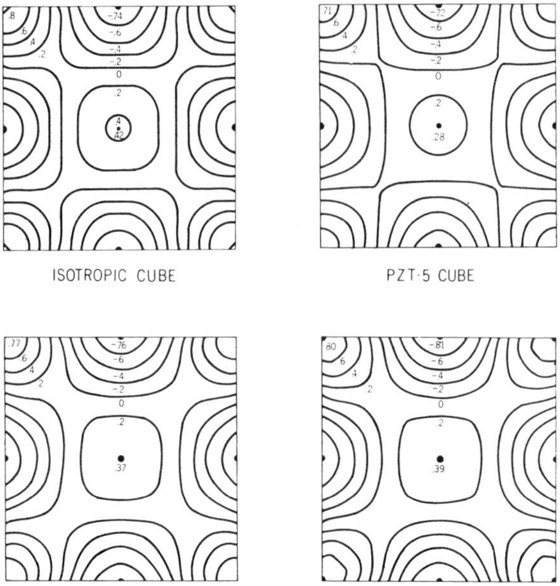

Figure 4.10. Level curves of $\psi_3^{(\nu)}$ at $x_3 = 0$ for the fourth in-phase dilation mode; $\nu_d = 3$ for the isotropic sample and $\nu_e = 4$ for the PZT-5 samples, $R_3 = \tau_3/\tau_1$.

For modes characterized by a large C_ν, the sample admittance in the vicinity of ω_ν will be high. This is usually considered to mean that the resonance at ω_ν is strong. In other words, the size of C_ν may be interpreted as a measure of the piezoelectric strength of mode ν.

However, a more convenient measure of this strength is C_ν normalized by the total dynamic capacitance $C_T - C_S$,

$$C_{\nu N} = \frac{C_\nu}{C_T - C_S} = \frac{C_\nu}{\sum_\nu C_\nu} \tag{4.104}$$

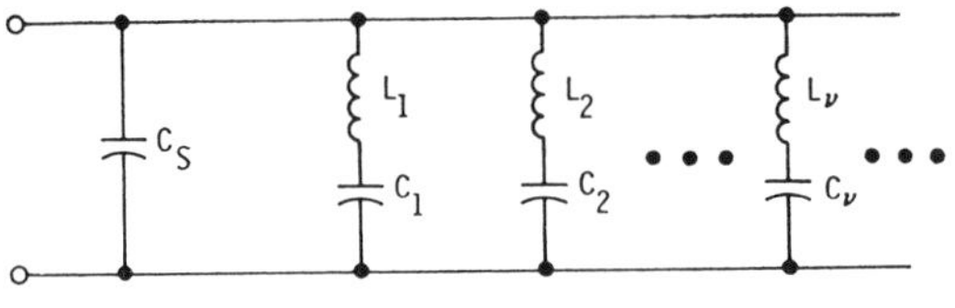

Figure 4.11. Equivalent circuit for the rectangular ferroelectric ceramic parallelepiped shown in Figure 4.3 (with external short circuit removed).

We shall refer to this $C_{\nu N}$ as the normalized dynamic capacitance of mode ν. The $C_{\nu N}$, of course, will be dimensionless, and their sum will be 1 because of Equation 4.104.

If we compare the equivalent circuit of Figure 4.11 with the formula for Y as given by Equation 4.98, we see that the dynamic capacitance of mode ν must be

$$C_\nu = \frac{(Q^{(\nu)})^2}{\omega_\nu^2} \tag{4.105}$$

and thus $C_{\nu N}$ must be

$$C_{\nu N} = \frac{(Q^{(\nu)})^2}{\omega_\nu^2(C_T - C_S)} \tag{4.106}$$

On the other hand, $Q^{(\nu)}$ as given by Equation 4.36 for the present geometry is

$$Q^{(\nu)} = -\int_0^{\tau_1}\int_0^{\tau_2} [e_{3kl}u_{k,l}^{(\nu)} - \epsilon_{3n}^S\phi_{,n}^{(\nu)}]^{x_3=\tau_3}\, dx_1\, dx_2 \tag{4.107}$$

Consequently, once Equations 4.89 and 4.90 have been solved for $u_i^{(\nu)}$ and $\phi^{(\nu)}$ and the solutions normalized, it should be possible to evaluate $C_{\nu N}$, the normalized dynamic capacitances, from Equations 4.106 and 4.107.

However, this is more complicated than one might, at first, think. Certain convergence problems are usually encountered when evaluating Equation 4.107 by direct substitution of Equations 4.89. In particular, the series for $u_i^{(\nu)}$ and $\phi^{(\nu)}$ are rather weakly convergent, and when differentiated as in Equation 4.107, may actually diverge. Fortunately, there is a simple solution to this difficulty.

The $Q^{(\nu)}$ of Equation 4.107 represents the total electric flux of mode ν evaluated at the top electrode $x_3 = \tau_3$. However, because of the zero divergence theorem, this total electric flux is actually independent of x_3. Consequently, it may be taken at any value of x_3 between 0 and τ_3, rather than being restricted to $x_3 = \tau_3$ as Equation 4.107 indicates. We may thus let $Q^{(\nu)}$ be expressed as the total electric flux averaged over x_3 between 0 and τ_3,

$$\begin{aligned} Q^{(\nu)} &= \frac{1}{\tau_3}\int_0^{\tau_3}\left[-\int_0^{\tau_1}\int_0^{\tau_2}(e_{3kl}u_{k,l}^{(\nu)} - \epsilon_{3n}^S\phi_{,n}^{(\nu)})\, dx_1\, dx_2\right] dx_3 \\ &= -\frac{1}{\tau_3}\iiint_V (e_{3kl}u_{k,l}^{(\nu)} - \epsilon_{3n}^S\phi_{,n}^{(\nu)})\, dV \end{aligned} \tag{4.108}$$

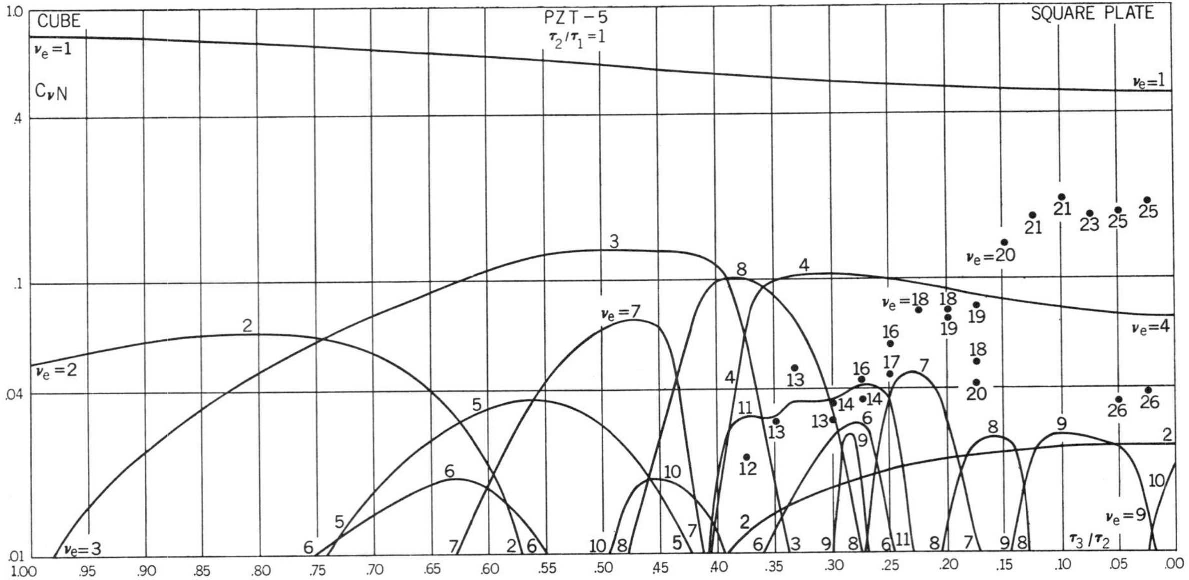

Figure 4.12. Normalized dynamic capacitances of the dilation family modes for a lead-zirconate-titanate ceramic (PZT-5) square cylinder. The thickness-to-width ratio is varied from unity (cube case) to zero (thin square plate case). The irregular series of points rising toward the right represents the thickness resonance complex.

This operation takes advantage of the fact that term by term integration improves the convergence of a series. It develops that this integration effectively neutralizes the deterioration in convergence introduced by the differentiation. Thus no difficulties are encountered when evaluating $C_{\nu N}$ by means of Equations 4.106 and 4.108.

Results of substituting the trial functions of Tables 4.2 and 4.3 into Equations 4.89, 4.106, and 4.108 are described by formulas tabulated in a supplementary report.[102] These formulas have been evaluated numerically by means of a digital computer program to find the normalized dynamic capacitances associated with the square cylinder spectrum of Figure 4.4, for PZT-5. Results are shown in Figure 4.12. It may be seen that only in-phase modes are piezoactive; as in the case discussed in Section 3.3, any contribution to $C_{\nu N}$ from some particle motion in the out-of-phase modes will be exactly cancelled by some corresponding opposite motion.

4.4.3. Application

One of the most troublesome problems in the construction of actual piezoelectric devices is that of obtaining clean thickness resonances in thin plates. These resonances are almost always plagued by spurious adjacent modes. It has long been known that the high contour-extensional overtones and contour-extensional-thickness hybrid modes make the spectrum very dense in this region.[107] While it is not possible to avoid these spurious resonances completely, it should be possible to find sample shapes and electrode configurations for which their strength as measured by dynamic capacitances is greatly suppressed.

Considerable insight into this problem may be gained from Figure 4.12. The irregular series of points rising toward the right of that figure represent the thickness resonance. It may, first of all, be seen that the thickness resonance is actually a composite of many high-order modes; the smaller the τ_3/τ_1 ratio, the higher the modal order. While these dynamic capacitances will peak and fall back as continuous functions when τ_3/τ_1 is decreased, the actual fluctuations are so rapid that they could not be drawn in this figure, except as discrete points, without great loss of clarity.

A possible approach to spurious mode suppression is well illustrated by comparing the situations that exist at $\tau_3/\tau_1 = 0.175$ and at $\tau_3/\tau_1 = 0.100$. At 0.175, the 18th, 19th, and 20th in-phase modes all have commensurate dynamic capacitances. Experimentally, this situation would manifest itself as a set of three closely spaced resonances, rather than a single clean resonance at the fundamental thickness frequency.

On the other hand, at 0.100 the 21st mode occurs alone and should give rise to a relatively clean resonance.

An alternative approach to spurious mode suppression is the energy-trapping theory presented in the next chapter. The energy-trapping method has gained wide commercial acceptance, which cannot be said of the present idea involving very careful choice of dimensional ratios.

Experimental verifications of the numerical results presented in this section have been published elsewhere,[108] along with additional numerical data.

4.5. Applications to Thick Piezoelectric Disks

4.5.1. Theory

We shall now conclude our variational work by evaluating the resonant properties of a thick, homogeneous ferroelectric ceramic disk poled in the axial direction. We shall discuss the case in which two thin concentric electrodes are present on the top surface and a full, grounded electrode is present on the bottom surface. This configuration is illustrated in Figure 4.13, along with dimensions, an electrode index convention, and a coordinate system.

The same problem was treated in Section 3.2 in the absence of piezoelectric effects. One of our present goals will be to see how that earlier purely elastic solution is modified by the inclusion of piezoelectricity.

As in the previous section dealing with piezoelectric parallelepipeds, the first step here is to determine the short-circuit natural modes. Consequently, we again begin by writing the short-circuit Lagrangian

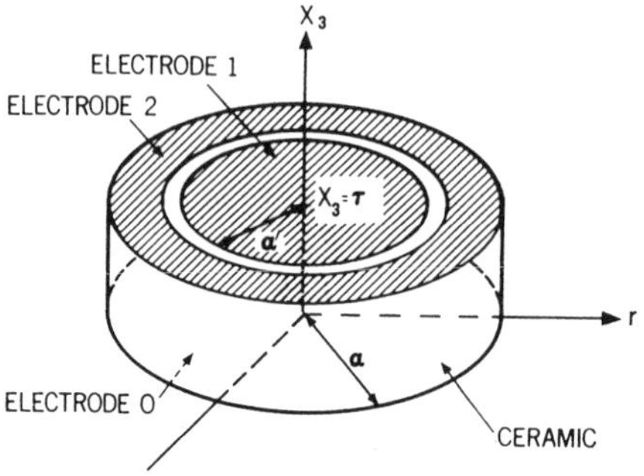

Figure 4.13. A symmetric three-electrode ferroelectric ceramic disk with comparable thickness and diameter.

of Equation 4.65 as specialized to the geometry in question:

$$L = \iiint_V \tfrac{1}{2}(u_{i,j}c^E_{ijkl}u_{k,l} + 2\phi_{,m}e_{mkl}u_{k,l} - \phi_{,m}\epsilon^S_{mn}\phi_{,n} - \rho\omega^2 u_i u_i)\, dV$$
$$- \iint_A [\phi(e_{3kl}u_{k,l} - \phi_{,n}\epsilon^S_{3n})]^{x_3=\tau}_{x_3=0}\, dA \qquad (4.109)$$

In this equation, A is the sample cross section in the $r\theta$-plane. Electrode volume $V - V'$ is assumed to be insignificant. Derivatives appearing in Equation 4.109 are covariant, and this is an important point here, where we deal with polar coordinates.

The u_i and ϕ are now expanded in trial function series (see Equations 4.66) to obtain an approximation for L corresponding to Equation 4.67. Operations 4.68 and 4.69 then yield the νth short-circuit mode of the disk,

$$u_i^{(\nu)} = \sum_{\alpha=1}^{S_m} B^{(\alpha,\nu)} U_i^{(\alpha)}$$
$$\phi^{(\nu)} = \sum_{\beta=1}^{S_e} C^{(\beta,\nu)} \Phi^{(\beta)} \qquad (4.110)$$

where the arrays $[B^{(\nu)}]$ and $[C^{(\nu)}]$ are the νth solutions of Equations 4.70,

$$[C^{(\nu)}] = [E^{-1}K_t][B^{(\nu)}]$$
$$[P + KE^{-1}K_t - \omega_\nu^2 H][B^{(\nu)}] = 0 \qquad (4.111)$$

The $[B^{(\nu)}]$ appearing in Equations 4.110 and 4.111, of course, must still obey the normalization condition of Equation 4.72.

Numerical studies of this problem have been carried out using the same expansion for u_i as in Section 3.2 (Equations 3.21 and 3.24). These trial functions all possess dilational symmetry and infinite rotational symmetry. We shall restrict our attention to modes of those symmetries, even though rotationally symmetric flexure or "oil-can" modes may also be excited by applying voltages to the electrodes in Figure 4.13.

Electric potential trial functions that have been found to give satisfactory results for the dilation modes are

$$\Phi^{(\beta)} = J_0(l_\beta r/a) \sin [2q_\beta \pi x_3/\tau] \qquad \beta = 1, 2, 3, \ldots, S_5$$
$$\Phi^{(\beta)} = \sin [2q_\beta \pi x_3/\tau] \qquad \beta = S_5 + 1, \quad S_5 + 2, \ldots, S_e \qquad (4.112)$$

Here l_β are the roots of

$$J_0'(l_\beta) = 0 \tag{4.113}$$

This set of trial functions, unlike others we have found to give good convergence in Chapters 3 and 4, is composed of a normwise complete expansion which is also pointwise complete but includes no additional over-complete correction.* It should be pointed out that these electric potential trial functions possess the dilational modal solution symmetries,

$$\Phi^{(\beta)} \quad \text{are odd in } x_3 \text{ with respect to the center of } V \text{ and } \theta \text{ independent} \tag{4.114}$$

For simplicity, we shall restrict our attention to the special case of negligible spacing between the two electrodes on the top surface in Figure 4.13. Then the short-circuit electrical boundary conditions on the top and bottom surfaces are

$$\phi\big|_{x_3=0,\tau} = 0 \tag{4.115}$$

We can see that this condition is obeyed by all the elements of Equation 4.112. While there is no requirement imposed by our theory to choose the $\Phi^{(\beta)}$ in this way, we can expect the benefit of accelerated convergence to result. Moreover, this choice permits us to omit the surface integral terms when computing the K and E matrices from Table 4.1.

When the formulas for H, P, K, and E appearing in Table 4.1 are evaluated in polar coordinates for a ferroelectric ceramic with appropriate taking of covariant derivatives, we find

$$\begin{aligned} K(\alpha,\beta) = \iiint\limits_V \Bigg[& \frac{\partial \Phi^{(\beta)}}{\partial x_3} e_{311} \left(\frac{\partial U_r^{(\alpha)}}{\partial r} + \frac{U_r^{(\alpha)}}{r} \right) + \frac{\partial \Phi^{(\beta)}}{\partial r} e_{113} \frac{\partial U_r^{(\alpha)}}{\partial x_3} \\ & + \frac{\partial \Phi^{(\beta)}}{\partial x_3} e_{333} \frac{\partial U_3^{(\alpha)}}{\partial x_3} + \frac{\partial \Phi^{(\beta)}}{\partial r} e_{113} \frac{\partial U_3^{(\alpha)}}{\partial r} \Bigg] dV \end{aligned} \tag{4.116}$$

$$E(\alpha,\beta) = \iiint\limits_V \left[\frac{\partial \Phi^{(\alpha)}}{\partial r} \epsilon_{11}^S \frac{\partial \Phi^{(\beta)}}{\partial r} + \frac{\partial \Phi^{(\alpha)}}{\partial x_3} \epsilon_{33}^S \frac{\partial \Phi^{(\beta)}}{\partial x_3} \right] dV$$

In addition, $P(\alpha,\beta)$ and $H(\alpha,\beta)$ are unchanged from Equations 3.22 and 3.23, except that c_{ijkl} must be replaced by c_{ijkl}^E. In obtaining Equation

* The sin $[2q_\beta \pi x_3/\tau]$ terms in Equations 4.112 which, at first, appear to constitute an over-complete correction actually just correspond to the $l_\beta = 0$ solution of Equation 4.113. Thus these terms merely complete the $J_0(l_\beta r/a)$ sin $[2q_\beta \pi x_3/\tau]$ expansion and do not make the resultant overcomplete.

4.116 from Table 4.1, surface integral contributions are omitted for the reasons discussed in the previous paragraph.

Let us now substitute the trial functions of Equations 3.21 and 4.112 into the matrix formulas of Equations 3.22, 3.23, and 4.116. This substitution gives recursion formulas for the sixteen submatrices of P and H, the eight submatrices of K, and the four submatrices of E. (The various submatrices represent all possible interactions of the six trial-function types in use here.) The resulting recursion formulas have been published elsewhere.[78]

Accuracy of 3 percent or so for ω_ν ($\nu < 10$) has been achieved with the 30 u_i trial functions of Equations 3.24, and with 22 trial functions for ϕ. In the notation of Equations 4.112 and 4.113, the 22 relevant ϕ trial functions are given by

$$\begin{aligned} S_5 = 16 \quad \text{with} \quad (l_\beta,q_\beta) = {} & (l_1,1), (l_2,1), (l_3,1), (l_4,1), \\ & (l_1,2), (l_2,2), (l_3,2), (l_4,2), \\ & (l_1,3), (l_2,3), (l_3,3), (l_4,3), \\ & (l_1,4), (l_2,4), (l_3,4), (l_4,4) \\ S_e - S_5 = 6 \quad \text{with} \quad q_\beta = {} & 1, 2, 3, 4, 5, 6 \end{aligned} \tag{4.117}$$

Upon computing ω_ν by inserting these trial functions into the characteristic value Equation 4.111, we will get the spectrum for this problem which is shown by the solid lines in Figure 4.14. These curves are based on the electroelastic coefficients listed in Table 4.4, which pertain to Clevite Ceramic A* ($BaTiO_3$). The dashed curves in Figure 4.14 show for comparison the results of Section 3.2 (Figure 3.2) where an identical computation was performed, except that piezoelectric effects were neglected. Also shown in Figure 4.14 is Shaw's experimental data for Ceramic A disks.[79] One may observe in Figure 4.14 that the inclusion of piezoelectricity raises the theoretical ω_ν values slightly and brings them into better agreement with the measurements.

All resonant frequency values shown in Figure 4.14 were multiplied by the disk thickness τ before being plotted. The horizontal axis in this figure represents the only remaining free parameter,

$$R = 2a/\tau \tag{4.118}$$

* Clevite Corporation, trade name.

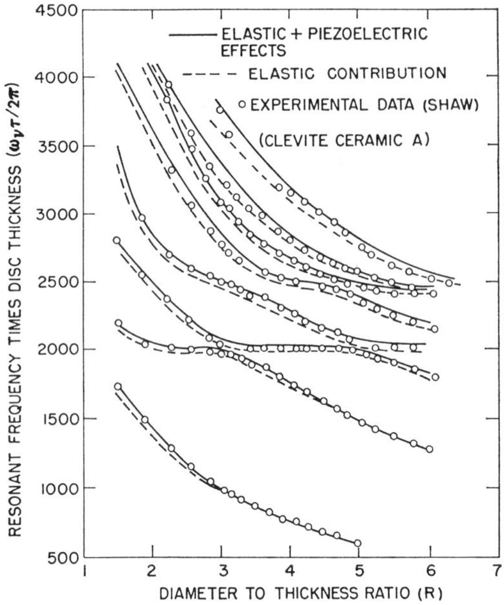

Figure 4.14. Comparison of theoretical and experimental resonant frequencies of thick ferroelectric ceramic disks. Solid lines show inclusion of piezoelectric effects, and dashed lines show their omission. Note that piezoelectric effects raise the resonant frequencies slightly.

the diameter-to-thickness ratio. Consequently, the spectrum shown in Figure 4.14 is representative of the short-circuit resonant frequencies of any disk characterized by the coefficients of Table 4.4, irrespective of absolute dimensions.

4.5.2. Determination of the Dynamic Capacitances

Let us first consider the case where a single electrode is present on the top of the disk, $a = a'$ in Figure 4.13. Then Equation 4.40 for the

Table 4.4. Electroelastic Coefficients and Density of Clevite Ceramic A ($BaTiO_3$) in MKS Units[17]

$c^E_{1111} = 1.50 \times 10^{11}$	$e_{311} = -4.35$
$c^E_{1122} = 0.66 \times 10^{11}$	$e_{333} = 17.5$
$c^E_{1133} = 0.66 \times 10^{11}$	$e_{113} = 11.4$
$c^E_{3333} = 1.46 \times 10^{11}$	$\epsilon^S_{11} = 9.9 \times 10^{-9} = 1115\epsilon_0$
$c_{1212} = 0.43 \times 10^{11}$	$\epsilon^S_{33} = 11.2 \times 10^{-9} = 1260\epsilon_0$
$c^E_{1313} = 0.44 \times 10^{11}$	$\rho = 5.7 \times 10^3$

admittance of that electrode becomes

$$Y_{11} = j\omega C_{11} + j\omega \sum_{\nu} \frac{(Q_1^{(\nu)})^2}{\omega_\nu^2 - \omega^2} \tag{4.119}$$

In this equation, C_{11} represents the clamped capacitance of the disk

$$C_{11} = \frac{\epsilon_{33}^S \pi a^2}{\tau} \tag{4.120}$$

This quantity corresponds to the parallelepiped C_S of Equation 4.99.

As with the rectangular parallelepiped, this single electrode-pair disk may be represented over all frequencies by the equivalent circuit shown in Figure 4.11 with $C_{11} = C_S$. The dynamic capacitance associated with the thick disk mode ν, C_ν, is obtained by comparing Equation 4.119 and Figure 4.11:

$$C_\nu = \frac{(Q_1^{(\nu)})^2}{\omega_\nu^2} \tag{4.121}$$

As in the previous section, the sum of all the dynamic capacitances C_ν in the single electrode case must add up to the difference between the zero-frequency capacitance and C_S,

$$C_T - C_S = \sum_{\nu} C_\nu \tag{4.122}$$

Consequently, the normalized dynamic capacitance,

$$C_{\nu N} = \frac{C_\nu}{\sum_{\nu} C_\nu} = \frac{(Q_1^{(\nu)})^2}{\omega_\nu^2(C_T - C_S)} \tag{4.123}$$

is again an excellent way of characterizing the piezoelectric strength of resonance ν. To compute this quantity for mode ν, we may determine ω_ν from Figure 4.14, $C_T - C_S$ from

$$C_T - C_S = (\epsilon_{33}^T - \epsilon_{33}^S)\pi a^2/\tau \tag{4.124*}$$

and $Q_1^{(\nu)}$ from Equation 4.36,

$$Q_1^{(\nu)} = -\int_0^a\int_0^{2\pi} [e_{3kl} u_{k,l}^{(\nu)} - \epsilon_{3n}^S \phi_{,n}^{(\nu)}]^{x_3=\tau} r\, dr\, d\theta \tag{4.125}$$

* As in Equation 4.103, the $\epsilon_{33}^T - \epsilon_{33}^S$ is related to the eigen coupling factor $k^{(5)}$ by $[k^{(5)}]^2\epsilon_{33}^T = \epsilon_{33}^T - \epsilon_{33}^S$.

We may, however, encounter convergence difficulties when evaluating Equation 4.125 by direct substitution of the series of Equations 4.110. These series for $u_i^{(\nu)}$ and $Q^{(\nu)}$ are not strongly convergent and, as in the previous section, may actually diverge when differentiated for Equation 4.125. It is best if one uses the zero divergence theorem to show that the integral of Equation 4.125, which is the total electric flux of mode ν, is actually independent of x_3. Thus $Q_1^{(\nu)}$ may be alternatively written as the total electric flux of mode ν averaged over x_3 between 0 and τ

$$Q_1^{(\nu)} = -\frac{1}{\tau}\iiint\limits_V (e_{3kl}u_{k,l}^{(\nu)} - \epsilon_{3n}^S\phi_{,n}^{(\nu)})\,dV \tag{4.126}$$

This operation introduces an additional term by term integration of the series expression for $Q_1^{(\nu)}$ and proves to eliminate all convergence failures.

Normalized dynamic capacitances for a fully electroded thick disk as computed from Equations 4.123 and 4.126 are illustrated in Figure 4.15. Trial functions of Equations 3.24 and 4.117 were utilized here.

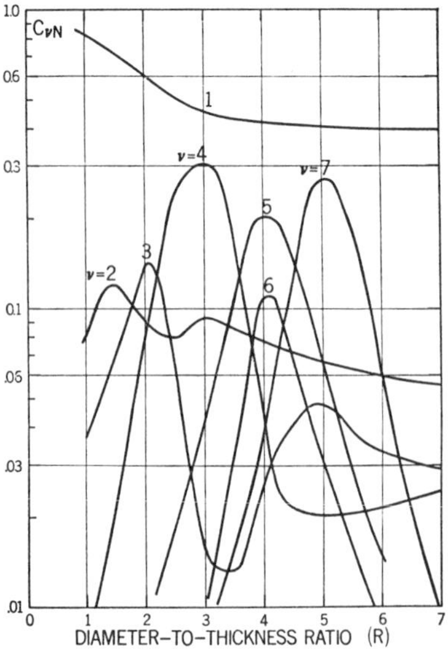

Figure 4.15. Normalized dynamic capacitances of the dilation modes in a thick barium titanate (Ceramic A) disk. The diameter-to-thickness ratio is plotted horizontally.

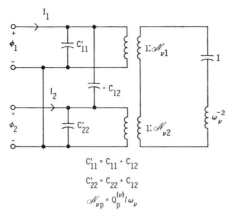

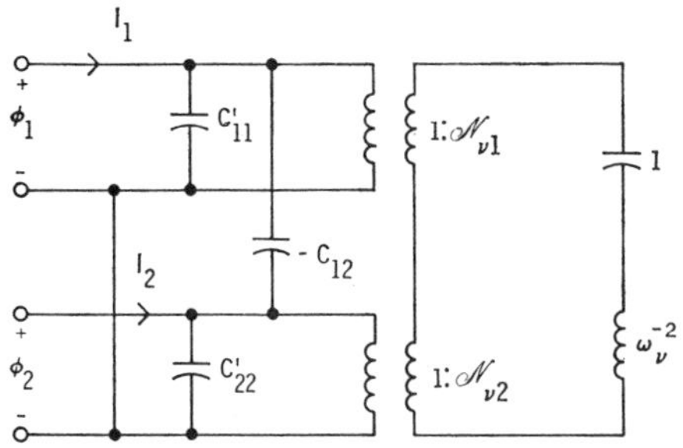

Figure 4.16. Equivalent circuit for the thick disk of Figure 4.13 near ω_ν.

Let us next consider the more complicated case shown in Figure 4.13 where two concentric electrodes are present on the top surface, and a very narrow electrode separation between them is located at $r = a'$. The equivalent circuit of Figure 4.2 then applies near ω_ν and may be specialized for application to the present situation as shown in Figure 4.16. In Figure 4.16, $C^{(\nu)}$ has been set equal to 1 because any one component of the transformer arrangement in Figure 4.2 may be assigned whatever value is convenient (see the discussion following Equation 4.48). Also, Figure 4.16 reflects the absence of mechanical terminals in the mechanically free disk under study here.

It may be seen in the equivalent circuit of Figure 4.16 that the important parameters now are ω_ν (which we know from Figure 4.14) and $Q_p^{(\nu)}/\omega_\nu$, where $p = 1, 2$. We should be able to evaluate $Q_p^{(\nu)}/\omega_\nu$ from Equation 4.36,

$$\frac{Q_p^{(\nu)}}{\omega_\nu} = -\frac{1}{\omega_\nu}\iint_{\partial V_p'} [e_{3kl}u_{k,l}^{(\nu)} - \epsilon_{3n}^S \phi_{,n}^{(\nu)}]\, dA \tag{4.127}$$

and Figure 4.14. However, as in Equation 4.125 for $Q_1^{(\nu)}$ in the fully electroded case, Equation 4.127 will usually exhibit convergence shortcomings when one tries to evaluate it by direct substitution of the series in Equations 4.110 for $u_i^{(\nu)}$ and $\phi^{(\nu)}$. It is again necessary to devise some way of introducing an integration into the computation to improve the convergence. This is most easily done by generalizing for this situation the technique of Equation 4.126; *i.e.*, the zero divergence theorem is used to justify an averaging integration.

In particular, let us first treat $Q_1^{(\nu)}$, the normalized free charge on the inner electrode at ω_ν. We define a closed Gaussian surface consisting of $x_3 = \tau, \tau'$, and $r = a'$, where $0 \leq \tau' \leq \tau$. The electric flux out of the $x_3 = \tau$ portion of the surface is $Q_1^{(\nu)}$, and this must equal the flux into the remaining portions. Since the exact value of this inward flux is independent of τ', its value as approximated by the series of Equations 4.110 and averaged over $0 \leq \tau' \leq \tau$ can be expected to be a reasonable estimate of the exact answer. Moreover, this averaging provides us with the desired excuse to include a convergence-aiding integration.

Explicitly, the averaged value for $Q_1^{(\nu)}$ is

$$Q_1^{(\nu)} = \frac{1}{\tau} \int_0^\tau \left[\int_{\tau'}^\tau -2\pi a' D_r^{(\nu)} \bigg|_{r=a'} dx_3 + \int_0^{a'} 2\pi r D_3^{(\nu)} \bigg|_{x_3=\tau'} dr \right] d\tau' \tag{4.128}$$

where $D_3^{(\nu)}$ and $D_r^{(\nu)}$ are obtained with covariant differentiation from Equations 4.2,

$$\begin{aligned} D_3^{(\nu)} &= \frac{\partial}{\partial x_3} \left(e_{333} u_3^{(\nu)} - \epsilon_{33}^S \phi^{(\nu)}\right) + e_{311} \left(\frac{\partial u_r^{(\nu)}}{\partial r} + \frac{u_r^{(\nu)}}{r}\right) \\ D_r^{(\nu)} &= e_{113} \left(\frac{\partial u_r^{(\nu)}}{\partial x_3} + \frac{\partial u_3^{(\nu)}}{\partial r}\right) - \epsilon_{11}^S \frac{\partial \phi^{(\nu)}}{\partial r} \end{aligned} \tag{4.129}$$

No convergence difficulties are encountered when $Q_1^{(\nu)}$ is calculated from Equations 4.128 and 4.129.

It is possible to obtain $Q_2^{(\nu)}$ analogously by setting up corresponding Gaussian surfaces. However, it is quicker merely to evaluate $Q_2^{(\nu)}$ from already known quantities,

$$Q_2^{(\nu)} = Q_1^{(\nu)}(a' = a) - Q_1^{(\nu)} \tag{4.130}$$

where $Q_1^{(\nu)}(a' = a)$ is the charge on electrode 1 in the single full-electrode case, as given by Equation 4.126.

Numerical values for $Q_p^{(\nu)}/\omega_\nu$, the transformer turns ratios in Figure 4.16, are most conveniently presented in dimensionless form analogously to $C_{\nu N}$ of Equation 4.123. In particular, let us divide $Q_p^{(\nu)}/\omega_\nu$ by $\sqrt{C_T - C_S}$ from the single full electrode case, Equation 4.122,

$$q_p^{(\nu)} = \frac{Q_p^{(\nu)}/\omega_\nu}{\sqrt{C_T - C_S}} \tag{4.131}$$

Equations 4.121 and 4.122 indicate that these $q_p^{(\nu)}$ have the desired

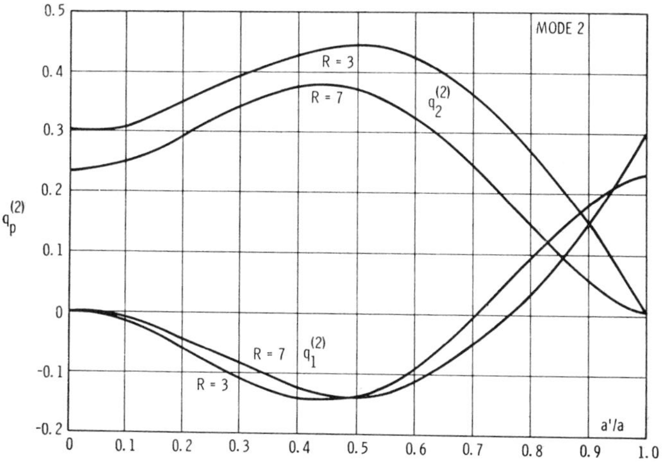

Figure 4.17. The normalized transformer turns ratio $q_p^{(2)}$ for mode 2 of a thick barium titanate disk (see Equation 4.133 and Figure 4.16; $R = 2a/\tau$).

dimensionless characteristic,

$$q_p^{(\nu)} = \frac{Q_p^{(\nu)}/\omega_\nu}{\left\{\sum_\nu [Q_1^{(\nu)}(a' = a)]^2/\omega_\nu^2\right\}^{1/2}} \tag{4.132}$$

On the other hand, if $q_p^{(\nu)}$ is known, we can quickly obtain the equivalent circuit transformer turns ratio $Q_p^{(\nu)}/\omega_\nu$ of Figure 4.16 from Equations 4.124 and 4.131:

$$Q_p^{(\nu)}/\omega_\nu = q_p^{(\nu)}[(\epsilon_{33}^T - \epsilon_{33}^S)\pi a^2/\tau]^{1/2} \tag{4.133}$$

Numerical values of $q_p^{(\nu)}$ for barium titanate disks, $\nu = 2, 3$, are shown in Figures 4.17 and 4.18 as functions of a'/a, inner electrode radius/disk radius. These curves are based on Equations 4.128 and 4.133, and the trial functions of Equations 3.24 and 4.117.

4.5.3. Device Applications

Thick ferroelectric ceramic disks with two or three concentric electrodes on one face have, generally speaking, the same applications as thin-plate multielectrode resonators (see Sections 2.1 and 2.2). For example, Figure 4.16 indicates that thick disks with two concentric electrodes may be used as tuned transformers. Also, it is possible to use thick disks with three concentric electrodes as the resonant elements

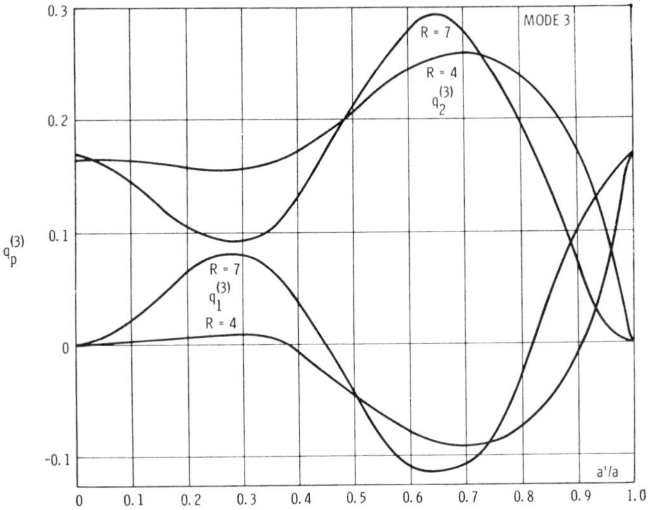

Figure 4.18. The normalized transformer turns ratio $q_p^{(3)}$ for mode 3 of a thick barium titanate disk (see Equation 4.133 and Figure 4.16; $R = 2a/\tau$).

in the electrically tuned oscillator circuit of Figure 2.4 or the FM discriminator circuit of Figure 2.6. In these devices, in fact, thick disks possess the obvious advantage of increased mechanical strength compared to thin plates. This may be a significant consideration for space or weapons program circuit designers: Applications that demand the radiation insensitivity of ferroelectric components usually also demand strong toleration to mechanical shock or acceleration.

It should be pointed out that for multistate information storage devices (see Figure 2.23) thick ferroelectric disks are not especially satisfactory. In this case, when attempting to read in polarization-information states, large signal fringing effects may present a real problem. Particularly if the electrodes are too close together, we will have difficulty changing the polarization state at one electrode pair without perturbing it seriously at adjacent electrode pairs. This situation is, of course, much less critical if we do not try to store too much information in a single disk. For example, we may use a small number of widely spaced electrode pairs and store binary rather than octal digits at each pair.

5 Energy Trapping

5.1. Introduction and Discussion of Device Applications

Around 1960, a major revolution in the design of high-frequency thickness-mode piezoelectric resonators and filters began to gain momentum. This revolution was made possible by a principle new to the field of acoustics — energy trapping.

The earliest phenomenological interpretations of acoustic energy trapping were based on an analogy with electromagnetic waveguides. Refer, for example, to the waveguide arrangement shown in Figure 5.1, where an enlarged waveguide section of width a is located between two narrower sections of width a'.

In the enlarged central section, no electromagnetic waves will propagate at frequencies below the cutoff of the first TE wave, $\omega_{co} = \pi c/a$. (Here, c is the speed of light.) However, at frequencies above ω_{co}, the first TE wave will propagate in the enlarged section. Similarly, in the narrower sections no wave will propagate below the cutoff frequency of the first TE wave in those sections $\omega'_{co} = \pi c/a'$. Note that ω'_{co} is a higher frequency than ω_{co}. Consequently, if somehow electromagnetic energy is injected into the central waveguide section at a frequency between ω_{co} and ω'_{co}, it will reflect back and forth in that section. However, it will be unable to escape into the narrower waveguide sections to the right or left. In other words, energy can be trapped in the central section of this arrangement if all possible escape routes have a cutoff frequency above that characterizing the excitation.

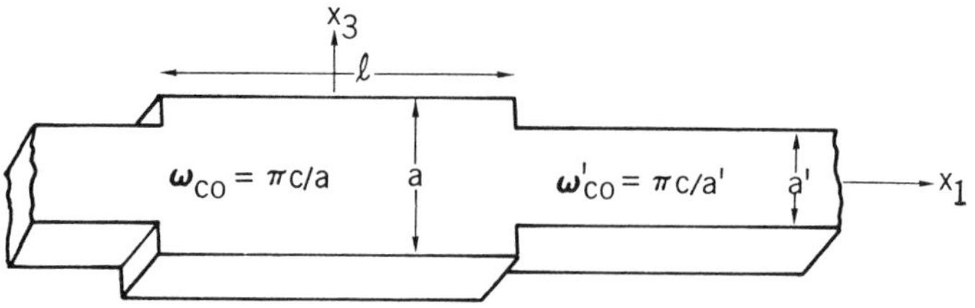

Figure 5.1. An enlarged electromagnetic waveguide section bounded by two narrower sections. For ω between ω_{co} and ω'_{co}, trapped waves can reflect back and forth in the enlarged section without penetrating into the narrower sections.

In the field of acoustics, this energy-trapping concept is utilized by constructing an elastic or piezoelectric plate consisting of one central region completely surrounded by a second outer region (see Figure 5.2). The two regions individually are specified to be uniform. Elastic and piezoelectric plates of finite thickness, like electromagnetic waveguides, support thickness-mode type waves. These waves also can either attenuate or propagate in the plane of the plate depending on whether the exciting frequency is above or below some cutoff frequency. Consequently, if the central region of the plate is fabricated so that its lowest thickness-wave cutoff frequency ω_{co} is below the corresponding frequency ω'_{co} of the outer region, it is possible to trap acoustic energy in the central region. To do this, the exciting frequency must, of course, be between ω_{co} and ω'_{co}.

The question arises as to how we may achieve this lowering of the fundamental thickness-wave acoustic cutoff frequency in the central region. Two methods are in common use, frequently in combination.

The first of these methods is in direct analogy to the electromagnetic

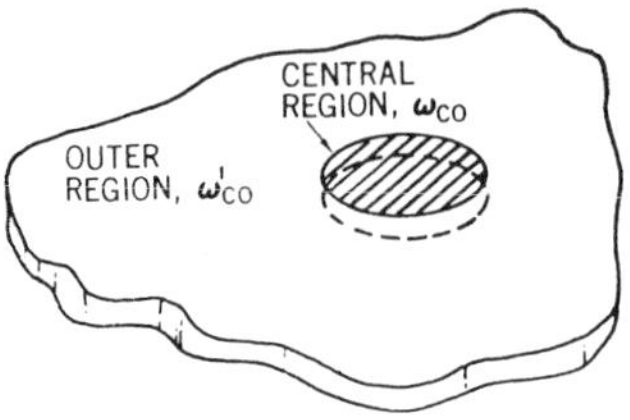

Figure 5.2. Basic acoustic energy-trapping configuration in a thin plate. The plate is so constructed that ω_{co}, the fundamental acoustic thickness-wave cutoff frequency in the central region, is lower than ω'_{co}, the corresponding outer region cutoff frequency.

case and will work either for piezoelectric or nonpiezoelectric plates: We need merely thicken the central region relative to the outer region, and ω_{co} will automatically be reduced below ω'_{co}. If the plate is piezoelectric, this thickening may be achieved by depositing electrodes on the desired region. These electrodes will then serve the dual purposes of lowering ω_{co} relative to ω'_{co} and of providing electrical access for injecting or withdrawing energy at the trapping region. For nonpiezoelectric plates, it is immaterial how the thickening is accomplished.

The second method for lowering ω_{co} is restricted in use to piezoelectric plates. This method operates on a much less intuitive basis than the first method. In particular, if infinitesimally thin electrodes are applied to the central region and not to the outer region, the desired lowering of ω_{co} will be achieved, even though the central region is not finitely thickened by the electrodes. The magnitude of this frequency lowering depends directly on the piezoactivity of the plate and inversely on the internal impedance of the electrical generator which excites the trapping region.* For a weakly active material such as quartz, the lowering may be only 0.3 percent or even less. On the other hand, for a strongly active material such as some ferroelectric ceramics, the lowering may exceed 20 percent.

Before we continue these introductory remarks on acoustic energy trapping, it is proper to record here a few historical observations. The electromagnetic waveguide analogue relationships described previously were published first by Mortley as early as 1951.[109,110] However, this remarkable work was so far ahead of its time that it was ignored for a number of years. In 1963, the analogue relationships were rediscovered by Shockley *et al.*[111] This time, technology had advanced to the point that many applications of energy trapping were immediately widely apparent. Great interest almost instantaneously greeted Shockley's work. The proliferation of devices and applications based on the principle of energy trapping which ensued in the years following 1963 is probably unparalleled in the field of ultrasonics. Then in 1967, Horton and Smythe[112] found Mortley's articles and informed the ultrasonic engineering community that this "new" technology was actually over a decade old.

However, before the wide acceptance of Shockley's work, and in some instances even before Mortley's publications, various phenomena later related to energy trapping had been observed or even utilized.

Let us, for instance, again refer to the electromagnetic waveguide

* See Section 5.4.

analogue shown in Figure 5.1. At some frequency slightly above ω_{co}, one standing wave can exist in a resonant condition along the dimension l in the x_1-direction. At a somewhat higher frequency, two resonant standing waves are permitted; at a still higher frequency, three; etc. Eventually, of course, frequency will exceed ω'_{co}, and then there can be no further standing wave resonances, as the energy escapes as fast as it is injected. The total number of permitted standing wave resonances of this type between ω_{co} and ω'_{co} is proportional to the length l of the enlarged region and to $\omega'_{co} - \omega_{co}$, the cutoff frequency separation in the two regions. Thus, if $l(\omega'_{co} - \omega_{co})$ is small enough, only the first resonance of this series may exist.

An identical situation exists in the case of piezoelectric trapped-energy resonators. However, in this case the resonances past the first show up as undesirable spurious responses. Prior to the theory proposed by Mortley or Shockley, Bechmann discovered that if electrodes on quartz plates were kept to small local areas, it was possible to obtain clean thickness mode resonators.[113,114] This small electrode condition, of course, is now known to be a generalization of the earlier-mentioned concept that a short waveguide section length l in Figure 5.1, and hence a small $l(\omega'_{co} - \omega_{co})$, will eliminate all the trapped waveguide resonances except the first.

Additional phenomena relating spurious responses to nonuniformities in plate thickness were observed by Gerber in 1944. These relations went unpublished until 1966 when they were interpreted as a consequence of energy trapping occurring at random local areas of the plate where it was slightly too thick.[115]

When energy-trapping devices first made their appearance, they primarily took the form of simple thickness mode piezoelectric resonators with single electrodes on the top and bottom of a plate. Virtually all early devices were fabricated from quartz. However, recently more sophisticated devices have been developed with two or more trapping regions located in close proximity to each other. It is then possible for energy to "tunnel" from one trapping region to another. These multiregion trapped-energy configurations have enough variable parameters that it is possible to use them for synthesizing virtually any desired filter-type response. For example, an entire lattice filter section with a single resonant circuit in each branch may be replaced by a single monolithic device consisting of two adjacent energy-trapping regions on a quartz plate.[116,117]

Still more complicated lattice filters with multiple resonant circuits in each branch may be synthesized by trapped-energy devices with more than two trapping regions.[118,119,120] In the case where more than two interacting trapping regions are present, one region usually serves as the electrical input, and one as the output. In the remaining intermediate regions, no piezoelectric interactions take place — all processes are entirely acoustical. Consequently, these intermediate regions unlike the input and output regions, need not even be provided with electrodes.[119]

Fundamental thickness mode quartz resonators, either conventional or energy-trapping, appear to have a practical upper frequency limitation around 25 MHz. This limit is imposed because it becomes very difficult to machine plates thinner than a half wavelength at that frequency. Consequently, above 25 MHz, thickness overtone resonators are used — usually the third or fifth overtone. In this way, the spectrum within reach of quartz resonators is extended to 125 MHz. Our reason for mentioning this fact here is that various engineers have recently succeeded in applying energy-trapping principles to these thickness-overtone type resonators.[121,122]

Especially in Japan and the Netherlands, interest has developed in ferroelectric ceramic, as well as quartz, trapped-energy devices. In this case, two principal classes of operation are possible: thickness twist or shear[123] and thickness dilation.[124] Our work in this chapter will consider ferroelectric ceramic resonators of both classes. First, we shall take up the twist-mode operation, because it is simpler, provides insight without overly complicating details, and can be treated either with or without some usually made approximations. Second, we shall discuss the dilation-mode operation, which is quite complicated analytically, even when all reasonable simplifying assumptions have been applied.

5.2. Thickness-Twist Modes

5.2.1. The Wave Equation

Let us now consider the phenomenon of energy trapping as it pertains to thickness-twist modes in ferroelectric ceramic plates. Figure 5.3 shows a section of an infinite plate uniformly polarized in its own plane (in the x_3-direction) with infinite strip electrodes on the top and bottom surfaces running parallel to polarization. Dimensions and coordinates to be used in this section are also illustrated in Figure 5.3.

We shall assume this problem is two-dimensional with no variations occurring along x_3. In contrast to the first four chapters of this monograph, we will now use $e^{-j\omega t}$, not $e^{+j\omega t}$ time dependence. (In the next

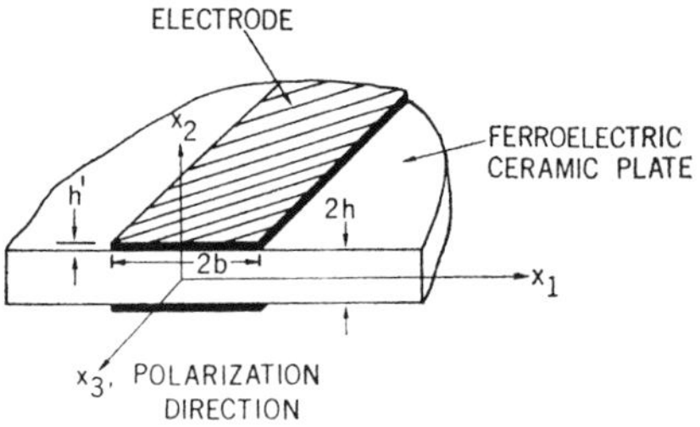

Figure 5.3. A thin ferroelectric ceramic plate polarized along x_3 with infinite strip electrodes along x_3.

section, this permits us to work with modified Hankel functions of the first kind, rather than the more complicated second kind.) This sign change also necessitates reversing the sign of the loss or imaginary components of the material coefficients in Equations 1.33 and 1.45: Power dissipation P_d in Equation 1.44 is not invariant to the transformation $e^{j\omega t} \to e^{-j\omega t}$ (i.e., $\omega \to -\omega$) unless this sign reversal in the material coefficients is made as well.

We shall search for thickness-twist type wave solutions of the equations of motion in which propagation is along x_1, u_3 and ϕ are nonzero, and u_1 and u_2 are zero.* The solutions we seek for the nonzero variables u_3 and ϕ will depend on x_1 and x_2. Subject to these restrictions, we find the nonzero strains are S_4 and S_5,

$$S_1 = \frac{\partial u_1}{\partial x_1} = 0 \qquad S_2 = \frac{\partial u_2}{\partial x_2} = 0 \qquad S_3 = \frac{\partial u_3}{\partial x_3} = 0$$

$$S_4 = \frac{\partial u_2}{\partial x_3} + \frac{\partial u_3}{\partial x_2} = \frac{\partial u_3}{\partial x_2} \qquad S_5 = \frac{\partial u_1}{\partial x_3} + \frac{\partial u_3}{\partial x_1} = \frac{\partial u_3}{\partial x_1} \tag{5.1}$$

$$S_6 = \frac{\partial u_1}{\partial x_2} + \frac{\partial u_2}{\partial x_1} = 0$$

Similarly, the nonzero electric fields are E_1 and E_2,

$$E_1 = -\frac{\partial \phi}{\partial x_1} \qquad E_2 = -\frac{\partial \phi}{\partial x_2} \qquad E_3 = -\frac{\partial \phi}{\partial x_3} = 0 \tag{5.2}$$

* Thickness-twist wave solutions resemble thickness-shear solutions in that both have particle displacement and propagation in the plane of the plate. However, in thickness-twist solutions, displacement and propagation are perpendicular, whereas they are parallel in thickness-shear solutions (see Figure 11 of Reference 111).

From the form of the constituent relations for a ferroelectric ceramic (class C_{6v} in Table 1.1), and from Equations 5.1 and 5.2, we can infer that T_4 and T_5 will be the only stresses and that D_1 and D_2 will be the only electric displacements in this configuration:

$$\begin{aligned}
T_4 &= c_{44}^E S_4 - e_{24} E_2 = c_{44}^E \frac{\partial u_3}{\partial x_2} + e_{24} \frac{\partial \phi}{\partial x_2} \\
T_5 &= c_{44}^E S_5 - e_{24} E_1 = c_{44}^E \frac{\partial u_3}{\partial x_1} + e_{24} \frac{\partial \phi}{\partial x_1} \\
D_1 &= e_{24} S_5 + \epsilon_{11}^S E_1 = e_{24} \frac{\partial u_3}{\partial x_1} - \epsilon_{11}^S \frac{\partial \phi}{\partial x_1} \\
D_2 &= e_{24} S_4 + \epsilon_{11}^S E_2 = e_{24} \frac{\partial u_3}{\partial x_2} - \epsilon_{11}^S \frac{\partial \phi}{\partial x_2}
\end{aligned} \tag{5.3}$$

Hence, conservation of momentum in differential form will yield

$$\begin{aligned}
\frac{\partial T_1}{\partial x_1} + \frac{\partial T_6}{\partial x_2} + \frac{\partial T_5}{\partial x_3} &= -\rho\omega^2 u_1 = 0 \\
\frac{\partial T_6}{\partial x_1} + \frac{\partial T_2}{\partial x_2} + \frac{\partial T_4}{\partial x_3} &= -\rho\omega^2 u_2 = 0 \\
\frac{\partial T_5}{\partial x_1} + \frac{\partial T_4}{\partial x_2} + \frac{\partial T_3}{\partial x_3} &= -\rho\omega^2 u_3 = c_{44}^E \nabla^2 u_3 + e_{24} \nabla^2 \phi
\end{aligned} \tag{5.4}$$

From these equations we see that there do, indeed, exist self-consistent thickness-twist solutions of the equations of motion depending only on x_1 and x_2 in which u_3 is the only particle displacement. The operator ∇^2 in Equation 5.4 is defined as

$$\nabla^2 = \frac{\partial^2}{\partial x_1^2} + \frac{\partial^2}{\partial x_2^2} \tag{5.5}$$

The condition that no free charge is permitted inside a piezoelectric insulator leads to

$$\frac{\partial D_1}{\partial x_1} + \frac{\partial D_2}{\partial x_2} = 0 = e_{24} \nabla^2 u_3 - \epsilon_{11}^S \nabla^2 \phi \tag{5.6}$$

Equations 5.4 and 5.6 may then be used to obtain the wave equation for u_3,

$$-\rho\omega^2 u_3 = (c_{44}^E + e_{24}^2/\epsilon_{11}^S)\nabla^2 u_3 = c_{44}^D \nabla^2 u_3 \tag{5.7}$$

where

$$\nabla^2\phi = \frac{e_{24}}{\epsilon_{11}^S}\nabla^2 u_3 \tag{5.8}$$

The constituent relations, Equations 1.26, have been used in simplifying the right side of Equation 5.7.

In the portion of the plate between the electrodes, the solution to Equations 5.7 and 5.8 which has the symmetry requisite for coupling to the electrodes is

$$\begin{aligned} u_3 &= A\cos \Xi x_1 \sin K x_2 \\ \phi &= \Phi(x_2)\cos \Xi x_1 \end{aligned} \tag{5.9}$$

This solution represents a standing wave along x_1, the postulated direction of propagation. Here, A is an arbitrary constant and Φ is an undetermined function of x_2. In addition, Ξ and K must obey

$$\Xi^2 + K^2 = \rho\omega^2/c_{44}^D \tag{5.10}$$

Substitution of Equations 5.9 into 5.8 indicates that Φ must satisfy the differential equation

$$\frac{d^2\Phi}{dx_2^2} - \Xi^2\Phi = -\frac{e_{24}}{\epsilon_{11}^S}(K^2 + \Xi^2)A\sin K x_2 \tag{5.11}$$

The general solution to Equation 5.11 is

$$\Phi = A\frac{e_{24}}{\epsilon_{11}^S}\sin K x_2 + B\sinh \Xi x_2 + C\cosh \Xi x_2 \tag{5.12}$$

Thus, the general x_1-standing-wave solution to Equations 5.7 and 5.8 in the electroded region that meets the symmetry requirements for coupling to the electrodes is

$$\begin{aligned} u_3 &= A\cos \Xi x_1 \sin K x_2 \\ \phi &= \left(A\frac{e_{24}}{\epsilon_{11}^S}\sin K x_2 + B\sinh \Xi x_2 + C\cosh \Xi x_2\right)\cos K x_1 \end{aligned} \tag{5.13}$$

where A, B, and C are undetermined coefficients.

In the electroded region, only solutions representing standing waves along x_1 are permitted. The situation outside the electrodes is just the opposite. There only traveling or evanescent waves may exist. With this understanding, repetition of steps 5.9 to 5.13 leads to the solution of

Equations 5.7 and 5.8 for u_3 and ϕ outside the electrodes,

$$u_3 = A' e^{j\Xi'(|x_1|-b)} \sin K' x_2$$

$$\phi = \left(A' \frac{e_{24}}{\epsilon_{11}^S} \sin K' x_2 + B' \sinh \Xi' x_2 + C' \cosh \Xi' x_2\right) \cdot e^{j\Xi'(|x_1|-b)} \tag{5.14}$$

Here A', B', and C' are additional undetermined coefficients and

$$\Xi'^2 + K'^2 = \rho\omega^2/c_{44}^D \tag{5.15}$$

5.2.2. Boundary Conditions and the Dispersion Relation in the Electroded Region

Let us first consider the boundary conditions in the electroded region. The most tractable solution to this problem is that resulting when the electrodes are both grounded and shorted together. This solution may be used later to compute the admittance of the electrode pair in accordance with the procedure described in Section 4.2. In this short-circuit case, the electrical boundary conditions become $\phi = 0$ at $x_2 = \pm h$. Application of that condition reduces the general solution, Equation 5.13, for u_3 and ϕ to

$$\begin{aligned} u_3 &= A \cos \Xi x_1 \sin K x_2 \\ \phi &= A \frac{e_{24}}{\epsilon_{11}^S}\left(\sin K x_2 - \sinh \Xi x_2 \frac{\sin Kh}{\sinh \Xi h}\right) \cos \Xi x_1 \end{aligned} \tag{5.16}$$

The mechanical boundary condition in the electroded region at $x_2 = \pm h$ is most simply treated by neglecting the stiffness of the electrode and including only inertia effects. This approximation is quite permissible because the electrodes are located in the low-stress regions at the plate surfaces. Hence, at $x_2 = h$, we find

$$\rho' h' \omega^2 u_3 = T_4 = c_{44}^E \frac{\partial u_3}{\partial x_2} + e_{24} \frac{\partial \phi}{\partial x_2} \tag{5.17}$$

Here, ρ' is the density of the electrode material. Substitution of Equation 5.16 in 5.17 and rearrangement then yields the thickness-twist dispersion relation between Ξ and ω in the electroded region,

$$\tan Kh = Kh\left([k^{(1)}]^2 \frac{\Xi h}{\tanh \Xi h} + R \frac{\omega^2 \rho h^2}{c_{44}^D}\right)^{-1} \tag{5.18}$$

In this equation, R is defined as

$$R = \rho' h' / \rho h \tag{5.19}$$

and $k^{(1)}$ is the eigen coupling factor of Equation 1.87,

$$[k^{(1)}]^2 = \frac{d_{15}^2}{\epsilon_{11}^S s_{44}^E} = \frac{e_{24}^2}{\epsilon_{11}^S c_{44}^D} \tag{5.20}$$

and liberal use has been made of the material coefficient identities, Equations 1.22, 1.24, and 1.26.

With the application of Equation 5.10, K may be eliminated from the dispersion relation:

$$\tan\left\{\frac{\rho\omega^2 h^2}{c_{44}^D} - \Xi^2 h^2\right\}^{1/2} = \left\{\frac{\rho\omega^2 h^2}{c_{44}^D} - \Xi^2 h^2\right\}^{1/2}\left([k^{(1)}]^2 \frac{\Xi h}{\tanh \Xi h} + R\,\frac{\omega^2 \rho h^2}{c_{44}^D}\right)^{-1} \tag{5.21}$$

Numerical solutions for the thickness-twist dispersion equation are best expressed in dimensionless form. To effect this, let us define the normalized frequency

$$\Omega^2 = \frac{\omega^2 h^2 \rho}{\pi^2 c_{44}^D} \tag{5.22}$$

and the normalized wavenumbers

$$\begin{aligned} L &= \Xi h \\ \mathscr{K} &= Kh = \{\pi^2\Omega^2 - L^2\}^{1/2} \end{aligned} \tag{5.23}$$

With these, Equation 5.21 becomes

$$\tan\{\pi^2\Omega^2 - L^2\}^{1/2} = \{\pi^2\Omega^2 - L^2\}^{1/2}\left(\frac{[k^{(1)}]^2 L}{\tanh L} + R\pi^2\Omega^2\right)^{-1} \tag{5.24}$$

The first five branches of L, designated $L_n(\Omega)$, for $n = 1$ to 5 have been computed from Equation 5.24 as functions of Ω for $0 \le \Omega \le 5$. This range of Ω includes all frequencies we would encounter in working up to and including the ninth thickness-shear overtone of the plate. Solutions for $L_n(\Omega)$ are illustrated in Figures 5.4 to 5.8 for the case of $R = 0$ (negligibly thin electroding) and $k^{(1)} = 0.685$. This value of $k^{(1)}$ pertains to the ferroelectric ceramic PZT-5, as is demonstrated by reference to Table 1.3 and Equation 5.20.

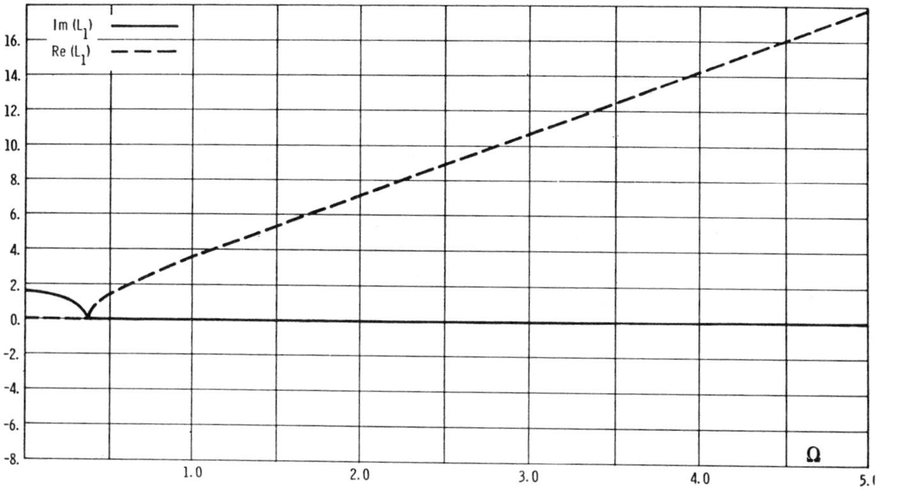

Figure 5.4. The first branch of the dispersion curve $L_1(\Omega)$ for thickness-twist waves in the electroded region of a PZT-5 ferroelectric ceramic plate.

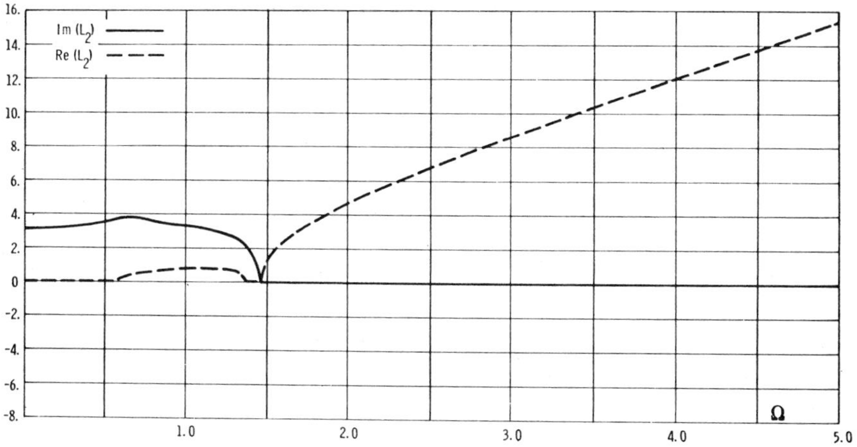

Figure 5.5. The second branch of the dispersion curve $L_2(\Omega)$ for thickness-twist in the electroded region of a PZT-5 ferroelectric ceramic plate.

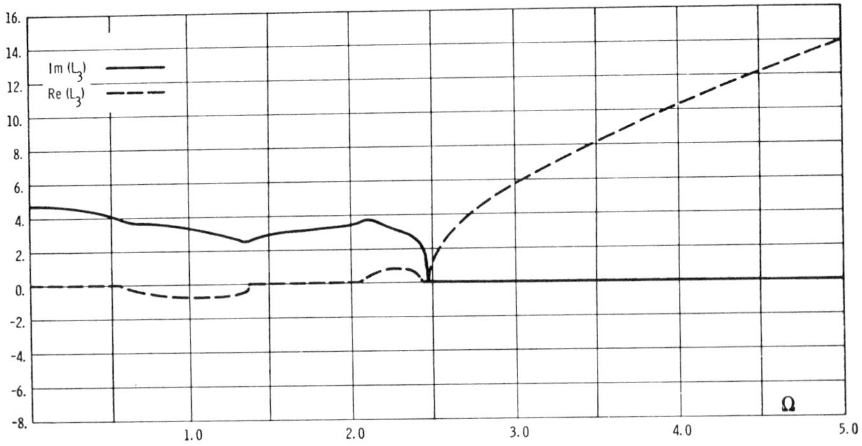

Figure 5.6. The third branch of the dispersion curve $L_3(\Omega)$ for thickness-twist waves in the electroded region of a PZT-5 ferroelectric ceramic plate.

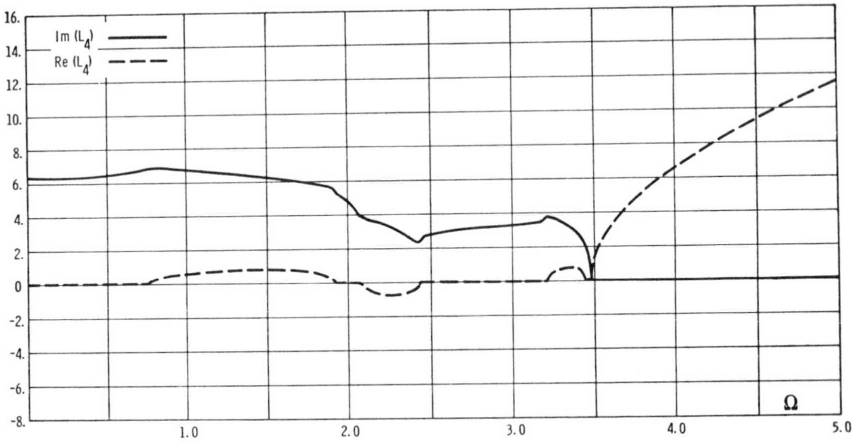

Figure 5.7. The fourth branch of the dispersion curve $L_4(\Omega)$ for thickness-twist waves in the electroded region of a PZT-5 ferroelectric ceramic plate.

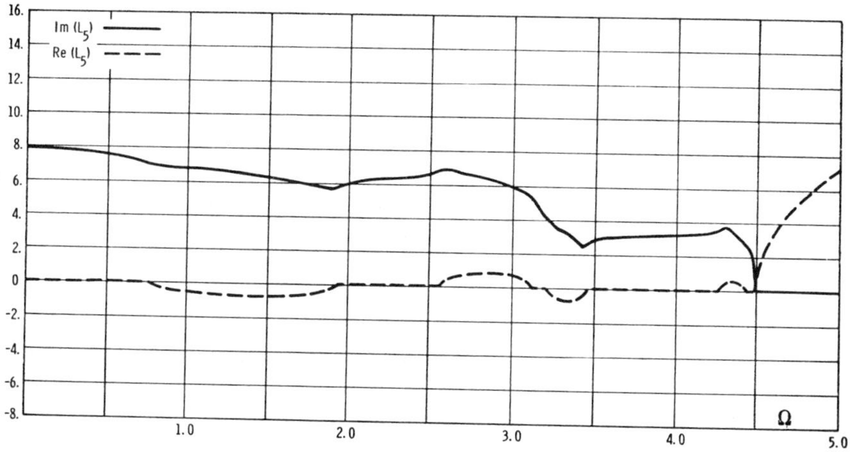

Figure 5.8. The fifth branch of the dispersion curve $L_5(\Omega)$ for thickness-twist waves in the electroded region of a PZT-5 ferroelectric ceramic plate.

For example, Figure 5.4 shows the first branch of the dispersion relation, $L_1(\Omega)$. This result is exactly the sort of curve one would anticipate: $L_1(\Omega)$ is imaginary below the cutoff frequency of $\Omega = 0.3811$ and real above $\Omega = 0.3811$. The special significance of the cutoff frequency $\Omega = 0.3811$ is that this Ω is the first solution of Equation 5.24 with L and R zero,

$$\tan \pi\Omega = \frac{\pi\Omega}{[k^{(1)}]^2} \tag{5.25}$$

Equation 5.25 is the familiar formula that governs the resonant frequencies of the thickness-shear modes in an infinite piezoelectric plate uniformly covered with thin electrodes. In other words, the cutoff frequency of the first thickness-twist wave in the electroded region is the same as the first thickness-shear resonant frequency of a completely electroded infinite plate.

Figure 5.5 shows $L_2(\Omega)$, the second branch of the dispersion curve. A rather unexpected development occurs here. Above the second solution of Equation 5.25, $\Omega = 1.4677$, $L_2(\Omega)$ is pure real. Also below $\Omega = 1.4677$, which is the cutoff frequency of the second symmetric thickness-twist traveling wave, $L_2(\Omega)$ has a positive imaginary component. However, between $\Omega = 0.56$ and 1.38, $L_2(\Omega)$ is complex rather than pure imaginary.

Figure 5.6 shows $L_3(\Omega)$. Again the wavenumber is pure real above its cutoff $\Omega = 2.4809$ and has a positive imaginary component below cutoff. However, there are now two regions below cutoff in which this branch of the dispersion curve is complex: $0.56 \le \Omega \le 1.38$ and $2.06 \le \Omega \le 2.43$.

The fourth and fifth branches of the dispersion curve are shown in Figures 5.7 and 5.8, respectively. Note that in each case there remains a well-defined cutoff frequency with $L_n(\Omega)$ pure real above and at least partly positive-imaginary below. However $L_4(\Omega)$ has three complex regions below cutoff, and $L_5(\Omega)$ has four. It is probably generally true for all n that $L_n(\Omega)$ has

1. a well-defined cutoff frequency asymptotically approaching $(n - \frac{1}{2})$;
2. only a real component above cutoff;
3. a positive imaginary component below cutoff; and
4. $n - 1$ complex regions below cutoff.

The first of these claims follows from Equation 5.25. No attempt has been made to verify the other three claims, although, based on Figures 5.4 to 5.8, their validity seems highly likely.

Let us conclude this subsection by defining the normalized x_2-wavenumber associated with L_n,

$$\mathscr{K}_n = \{\pi^2\Omega^2 - L_n^2\}^{1/2} \tag{5.26}$$

(see Equation 5.23); the unnormalized wavenumbers K and Ξ associated with L_n,

$$\begin{aligned} K_n &= \mathscr{K}_n/h \\ \Xi_n &= L_n/h \end{aligned} \tag{5.27}$$

(see Equation 5.23), the u_3 and ϕ associated with L_n,

$$u_{3(n)} = A_n \cos \Xi_n x_1 \sin K_n x_2 \tag{5.28}$$

$$\phi_{(n)} = A_n \frac{e_{24}}{\epsilon_{11}^S}\left(\sin K_n x_2 - \sinh \Xi_n x_2 \frac{\sin K_n h}{\sinh \Xi_n h}\right) \cos \Xi_n x_1 \tag{5.29}$$

(see Equation 5.16), and the T_5 associated with L_n,

$$\begin{aligned} T_{5(n)} = & -c_{44}^D A_n \Xi_n \sin \Xi_n x_1 \sin K_n x_2 \\ & + \frac{e_{24}^2}{\epsilon_{11}^S} A_n \Xi_n \sin \Xi_n x_1 \sinh \Xi_n x_2 \frac{\sin K_n h}{\sinh \Xi_n h} \end{aligned} \tag{5.30}$$

(see Equation 5.3).

5.2.3. Boundary Conditions and the Dispersion Relation in the Unelectroded Region

Outside the electroded portion of the plate, a considerably simpler dispersion relation and set of boundary conditions apply. Let us begin by writing Equations 5.3, the constituent relations, as

$$T_4 = c_{44}^E \frac{\partial u_3}{\partial x_2} - \frac{e_{24}}{\epsilon_{11}^S}\left(D_2 - e_{24}\frac{\partial u_3}{\partial x_2}\right) \tag{5.31}$$

In the unelectroded region at $x_2 = \pm h$, the normal component of stress and of electric displacement T_4 and D_2 must vanish. This reduces Equation 5.31 to simply

$$\frac{\partial u_3}{\partial x_2} = 0 \qquad \text{at } x_2 = \pm h \tag{5.32}$$

Substitution of u_3 from Equations 5.14 then gives

$$\cos K'h = 0 \tag{5.33}$$

which has the solutions

$$K'_{n'}h = (n' - \tfrac{1}{2})\pi \qquad n' = 1, 2, 3, \ldots \tag{5.34}$$

We may then obtain the thickness-twist dispersion relation for the unelectroded region from this last equation and Equation 5.15: the n'th branch is

$$\Xi'_{n'}h = \{\rho\omega^2h^2/c_{44}^D - (n' - \tfrac{1}{2})^2\pi^2\}^{1/2} \tag{5.35}$$

In analogy with Equations 5.23, it is useful to define dimensionlessly normalized wavenumbers in the unelectroded region,

$$\begin{aligned} L'_{n'} &= \Xi'_{n'}h = \pi\{\Omega^2 - (n' - \tfrac{1}{2})^2\}^{1/2} \\ \mathscr{K}'_{n'} &= K'_{n'}h = \pi(n' - \tfrac{1}{2}) \end{aligned} \tag{5.36}$$

The first of these relatively simple relations corresponds to the very complicated dispersion relation, Equation 5.24, of the electroded region. In the unelectroded case, the normalized wavenumber of the n'th branch of the dispersion relation $L'_{n'}$ is simply pure real above cutoff ($\Omega = n' - \tfrac{1}{2}$) and pure positive imaginary below. Complex regions do not exist in this case.

For subsequent uses, we shall need to evaluate the quantities T_5, ϕ, and u_3 which are associated with $K'_{n'}$ in the unelectroded region. To begin with, the thickness-twist solution for ϕ is given by Equations 5.14 in the unelectroded region. This general solution is adapted to the present case first by setting C' equal to zero. We do this because we are looking for solutions of the original problem that correspond to the situation in which the two electrode strips are shorted together and grounded. Subject to that condition, Equation 5.16 for ϕ applies in the electroded region, and thus we find

$$\phi = 0 \qquad \text{at } (x_1,x_2) = (\pm b,0) \tag{5.37}$$

However, the points $(x_1,x_2) = (\pm b,0)$ are on the boundary between the electroded and unelectroded regions. Hence, Equation 5.37 must also apply to the solution for ϕ in the unelectroded region. Reference to Equations 5.14 indicates that this boundary condition can be met only if $C' = 0$.

Also, B' must be set equal to zero in specializing Equations 5.14 to this situation, because at $x_2 = \pm h$

$$D_2 = e_{24}\frac{\partial u_3}{\partial x_2} - \epsilon_{11}^S\frac{\partial \phi}{\partial x_2} = 0 \tag{5.38}$$

Substitution of Equations 5.14 into 5.38 yields the result

$$-\epsilon_{11}^S B'\Xi' \cosh \Xi' h \cdot e^{j\Xi'(|x_1|-b)} = 0 \tag{5.39}$$

which can only be satisfied if B' itself is zero, as stated earlier.

Thus the n'th thickness-twist wave solution for ϕ outside the electroded region becomes

$$\phi_{(n')} = A'_{n'}\frac{e_{24}}{\epsilon_{11}^S}\sin\left[(n' - \tfrac{1}{2})\pi x_2/h\right]e^{j\Xi'_{n'}(|x_1|-b)} \tag{5.40}$$

The corresponding solution for u_3 is also obtained from Equations 5.14:

$$u_{3(n')} = A'_{n'}\sin\left[(n' - \tfrac{1}{2})\pi x_2/h\right]e^{j\Xi'_{n'}(|x_1|-b)} \tag{5.41}$$

Lastly, the associated expression for T_5 is found from Equations 5.3, 5.40, and 5.41:

$$T_{5(n')} = \pm A'_{n'}c_{44}^D j\Xi'_{n'}\sin\left[(n' - \tfrac{1}{2})\pi x_2/h\right]e^{j\Xi'_{n'}(|x_1|-b)} \tag{5.42}$$

where the $+$ sign applies when $x_1 \geq b$, and the $-$ sign applies when $x_1 \leq -b$.

5.2.4. Boundary Conditions Between the Electroded and Unelectroded Regions; the Resonant Frequency Equation

There are three boundary conditions that must be considered at the edges of the electrode: u_3, T_5, and ϕ must be continuous functions of x_1 for all values of x_2 at $x_1 = \pm b$.

The first two of these conditions may be satisfied exactly within our present framework. Continuity of ϕ is somewhat less easy to achieve mathematically. Difficulties arise for the same reason presented in the previous chapter when explicating Equations 4.27: Unlike u_3 and T_5, the ϕ is not merely a linear combination of the modal or wave solutions but an additional component is present as well.

The restriction imposed by dropping C' in the expansion for ϕ outside the electroded region (see Equation 5.37 and the subsequent discussion) already guarantees the continuity of ϕ with respect to x_1 at the boundary points located in the center of the plate $(x_1,x_2) = (\pm b,0)$. Except for this restriction, we shall not attempt to satisfy the boundary condition on ϕ at the electrode edge.

It is possible to express particle displacement, unlike potential, exactly as a sum of the modal wave solutions, Equations 5.28 and 5.41. In other words, we may write

$$\begin{aligned} u_3 &= \sum_{n=1}^{\infty} A_n \cos \Xi_n x_1 \sin K_n x_2 \qquad |x_1| \leq b \\ u_3 &= \sum_{n'=1}^{\infty} A'_{n'} \sin\left[(n' - \tfrac{1}{2})\pi x_2/h\right] e^{j\Xi'_{n'}(|x_1|-b)} \qquad |x_1| \geq b \end{aligned} \tag{5.43}$$

If we apply the condition that these two expressions must be equal at $x_1 = \pm b$, then multiply the equality by $\sin\left[(m' - \tfrac{1}{2})\pi x_2/h\right]$ and integrate from $x_2 = -h$ to $x_2 = +h$, we find

$$A'_{n'}h = \sum_{n=1}^{\infty} A_n \cos \Xi_n b \cdot h I_{n'n} \tag{5.44}$$

where

$$\begin{aligned} I_{n'n} &= \frac{1}{h}\int_{-h}^{h} \sin\left[(n' - \tfrac{1}{2})\pi x_2/h\right] \sin K_n x_2 \, dx_2 \\ &= -\frac{(-1)^{n'} \cos K_n h \cdot 2K_n h}{(n' - \tfrac{1}{2})^2\pi^2 - (K_n h)^2} \end{aligned} \tag{5.45}$$

Similarly, we may express T_5 exactly as a sum of the modal wave stresses, Equations 5.30 and 5.42,

$$\begin{aligned} T_5 &= \sum_{n=1}^{\infty} A_n \Big(-c_{44}^D \Xi_n \sin \Xi_n x_1 \sin K_n x_2 \\ &\qquad + \frac{e_{24}^2}{\epsilon_{11}^S} \Xi_n \sin \Xi_n x_1 \sinh \Xi_n x_2 \frac{\sin K_n h}{\sinh \Xi_n h} \Big) \qquad |x_1| \leq b \\ T_5 &= \pm \sum_{n'=1}^{\infty} A'_{n'} c_{44}^D j \Xi'_{n'} \sin\left[(n' - \tfrac{1}{2})\pi x_2/h\right] e^{j\Xi'_{n'}(|x_1|-b)} \qquad |x_1| \geq b \end{aligned} \tag{5.46}$$

Application of the condition that T_5 be continuous at $x_1 = \pm b$, multiplication of the resulting equality by sin $[(m' - \frac{1}{2})\pi x_2/h]$, and integration from $x_2 = -h$ to $x_2 = +h$ then gives

$$A'_{n'} j\Xi'_{n'}h = \sum_n A_n\Xi_n h \sin \Xi_n b\left(-I_{n'n} + J_{n'n}[k^{(1)}]^2 \frac{\sin K_n h}{\sinh \Xi_n h}\right) \tag{5.47}$$

where

$$\begin{aligned} J_{n'n} &= \frac{1}{h}\int_{-h}^{h} \sin [(n' - \tfrac{1}{2})\pi x_2/h] \sinh \Xi_n x_2 \, dx_2 \\ &= -\frac{(-1)^{n'} \cosh \Xi_n h \cdot 2\Xi_n h}{(n' - \frac{1}{2})^2\pi^2 + (\Xi_n h)^2} \end{aligned} \tag{5.48}$$

and where $[k^{(1)}]$ is defined by Equation 5.20.

Equations 5.45 and 5.47 may be rearranged and written in dimensionless form as

$$\begin{aligned} &A'_{n'} - \sum_{n=1}^{\infty} A_n I_{n'n} \cos L_n B = 0 \\ &A'_{n'} jL'_{n'} + \sum_{n=1}^{\infty} A_n\left(I_{n'n} - J_{n'n}[k^{(1)}]^2 \frac{\sin \mathscr{K}_n}{\sinh L_n}\right) L_n \sin L_n B = 0 \end{aligned} \tag{5.49}$$

where B is the electrode width to plate thickness ratio,

$$B = b/h \tag{5.50}$$

and where the corresponding dimensionless forms for $I_{n'n}$ and $J_{n'n}$ are

$$\begin{aligned} I_{n'n} &= \frac{-(-1)^{n'} \cos \mathscr{K}_n \cdot 2\mathscr{K}_n}{\mathscr{K}'^2_{n'} - \mathscr{K}^2_n} \\ J_{n'n} &= \frac{-(-1)^{n'} \cosh L_n \cdot 2L_n}{\mathscr{K}'^2_{n'} + L^2_n} \end{aligned} \tag{5.51}$$

In obtaining this result, use is made of Equations 5.27 and 5.36.

The system given by Equation 5.49 comprises an infinite set of homogeneous linear equations in an infinite number of unknowns. Practically speaking, however, we must truncate both the A_n and the $A'_{n'}$ series after some finite number of terms before attempting to find a solution. If N terms are kept in both series, we will obtain $2N$ linear homogeneous equations in $2N$ unknowns. It is then possible to compute approximate frequencies for the trapped-energy resonances as functions of the parameter B. This is done as follows: The system of $2N$ equations, in 5.49, will have nontrivial solutions for A_n and $A'_{n'}$ only when the determinant of the associated coefficient array vanishes. This will happen only for certain discrete values of Ω — the values of Ω which correspond to the desired resonant frequencies.

Consequently, we proceed by (1) picking some value of Ω; (2) computing the first N values of $L_n(\Omega)$ from Equation 5.24, $\mathcal{K}_n(\Omega)$ from Equation 5.26, $L'_{n'}(\Omega)$ from Equations 5.36, and $\mathcal{K}'_{n'}(\Omega)$ from Equations 5.36; (3) substituting these four wavenumbers in Equations 5.49 and 5.51; (4) checking to see if the resulting $2N \times 2N$ coefficient determinant of Equations 5.49 is zero; and (5) trying a new value of Ω if it is not zero.

It is not necessary to examine all values of Ω for a vanishing determinant; solutions are permitted only in certain bands. In particular, in order for a trapped mode to exist which is associated with the nth thickness-twist wave in the plate, $L_n(\Omega)$ as given by Equation 5.24 must be real and $L'_n(\Omega)$ as given by Equations 5.36 must be imaginary. Under these conditions, the nth thickness-twist wave can propagate under the electrode (Equation 5.28) and attenuate outside the electrode (Equation 5.41), as is required to have an energy-trapped resonance. In other words, the energy-trapped resonant frequencies associated with the nth thickness-twist wave must lie between the nth solution of Equation 5.25 and $n - \frac{1}{2}$ (Equations 5.36). Numerically, for PZT-5 as described in Table 1.3, this means resonances must be between

0.3811 and 0.5000 for $n = 1$
1.4677 and 1.5000 for $n = 2$
2.4809 and 2.5000 for $n = 3$
3.4864 and 3.5000 for $n = 4$
4.4894 and 4.5000 for $n = 5$
etc.

Numerical solutions for the resonant modes of a PZT-5 plate have been computed from Equations 5.49 by taking the first three terms in each series; *i.e.*, $N = 3$. This has been done for $R = 0$ (negligible electrode thickness) and $4 \leq B \leq 5$. In this range of B, it is found that two resonances are associated with the first thickness-twist wave $n = 1$. The resulting spectrum is plotted in Figure 5.9. Here, the vertical axis is expanded to show only the frequency band of interest; $0.3811 \leq \Omega \leq 0.5000$. The lower cutoff 0.3811 is represented by Ω_{co}, and the upper cutoff 0.5000 by Ω'_{co}. The actual ordinate in Figure 5.9 is

$(\Omega_{nm} - \Omega_{co})/(\Omega'_{co} - \Omega_{co})$, the fractional frequency separation of Ω_{nm} between the lower and upper cutoff. In this notation, Ω_{nm} represents the mth normalized resonant frequency solution associated with the nth thickness-twist wave.

It is interesting to compare the values of Ω_{nm} in Figure 5.9 with those obtained when only one term, the first, is taken from each series in Equations 5.49. The resulting alteration in $(\Omega_{nm} - \Omega_{co})/(\Omega'_{co} - \Omega_{co})$

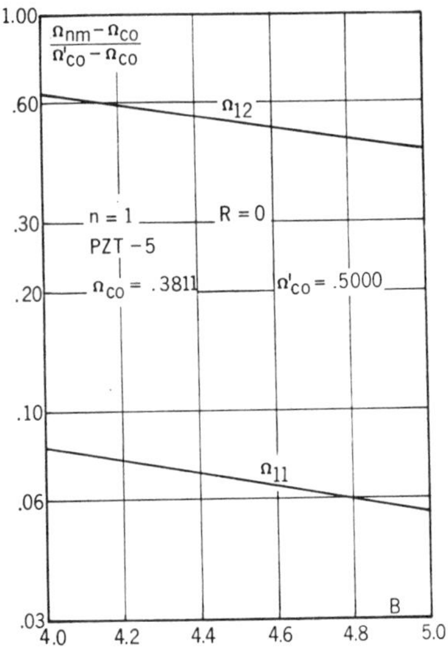

Figure 5.9. Trapped-energy resonance spectrum of the first thickness-twist wave for a PZT-5 plate. Values for Ω_{nm} were computed by coupling the first three thickness-twist wave solutions in the electroded and unelectroded regions.

is always found to be less than 1 percent! This is a particularly useful discovery, since the computational simplification in solving a 2×2 rather than a 6×6 determinant is tremendous.

Values for $(\Omega_{nm} - \Omega_{co})/(\Omega'_{co} - \Omega_{co})$ as computed with just the first term in each series of Equations 5.49 are plotted in Figure 5.10 for $R = 0$ and $0 \leq B \leq 10$. This figure also is based on the material coefficients of PZT-5. We can see that for $0 < B < 2.5$, only one resonance is present; for $2.5 < B < 5.0$, two resonances are present; for $5.0 < B < 7.5$, three resonances are present; and for $7.5 < B < 10.0$, four resonances are present. Consequently, if we wish to design a trapped-resonance PZT-5 filter of this type, we should select the dimensions so that $B < 2.5$ in order to exclude spurious responses.

Trapped-energy resonances are also associated with the higher thickness-twist waves. However, in this case a complicating effect occurs. In particular, at the electrode edge mode conversion can take place. This is not important in dealing with $n = 1$ trapped resonances, because in that case no waves propagate in the unelectroded region: It does not really matter if energy is scattered out of the $n = 1$ wave.

But if we are dealing, say, with an $n = 2$ trapped resonance, it is very important if energy is scattered into an $n = 1$ wave by the electrode edge. In this case, the $n = 1$ wave does not attenuate in the unelectroded region, and whatever energy is scattered into the $n = 1$ wave will leak

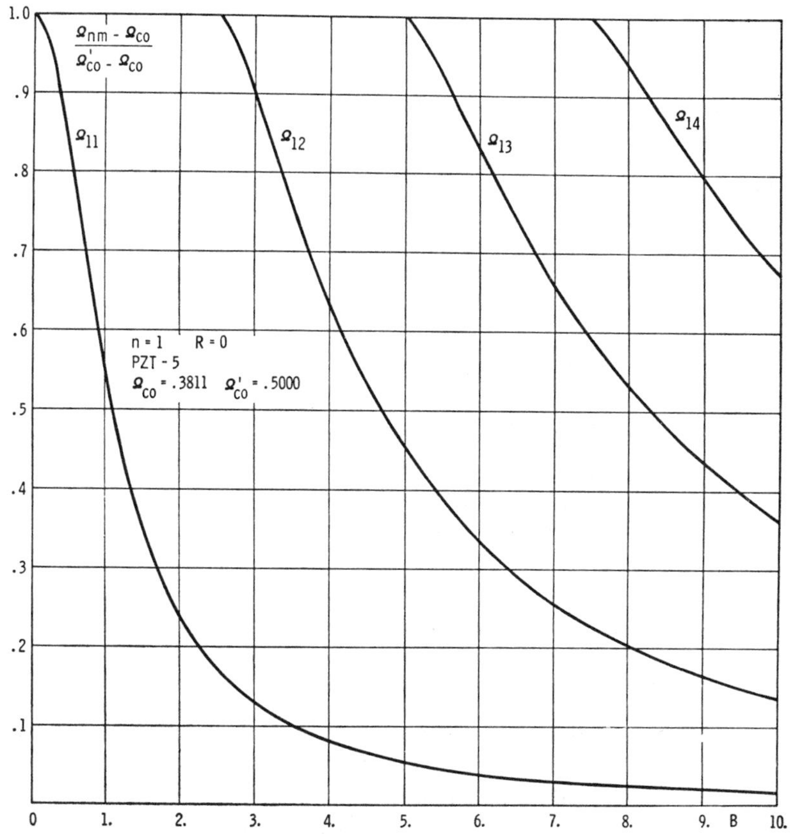

Figure 5.10. Trapped-energy resonance spectrum of the first thickness-twist wave for a PZT-5 plate. Values were computed by coupling the first thickness-twist wave solutions in the electroded and unelectroded regions.

away. Consequently, we may expect $n = 2$ trapped resonances to exhibit exponential time decay even in the absence of material dissipation; these resonances are not perfectly trapped. Mathematically, this means Ω_{nm} will be complex with a negative imaginary part if $n \geq 2$.

Indeed, the $n = 2$ solutions of Equations 5.49 prove to be complex. For the $n = 2$ thickness-twist wave, the electroded region cutoff

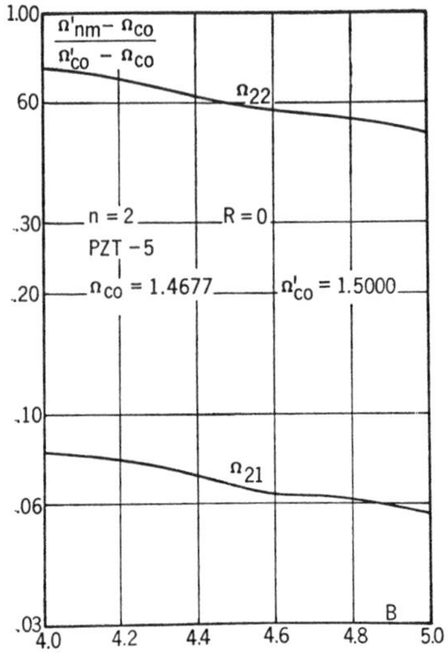

Figure 5.11. Trapped-energy resonance spectrum of the second thickness-twist wave for a PZT-5 plate. Values for Ω_{nm} were computed by coupling the first three thickness-twist wave solutions in the electroded and unelectroded regions. Only the real part of Ω_{nm} is shown here; the imaginary part is presented in Figure 5.12.

frequency is $\Omega_{co} = 1.4677$ and the unelectroded region cutoff frequency is $\Omega'_{co} = 1.5000$. If we then define

$$\Omega_{nm} = \Omega'_{nm} - j\Omega''_{nm} \tag{5.52}$$

we can conveniently plot the real part of Ω_{2m} as the fractional frequency difference of Ω'_{2m} between Ω_{co} and Ω'_{co}. This has been done in Figure 5.11 for PZT-5 in the case of $R = 0$. Three terms were taken in each series of Equations 5.49 to perform these computations; the $n = 1$ term corresponds to the leak. As is apparent from Figure 5.11, for $4 \leq B \leq 5$ there are two $n = 2$ trapped resonances, just as was the case with $n = 1$.

The imaginary part of Ω_{2m} may be presented in the form of a quality factor,

$$Q_{nm} = \Omega'_{nm}/(2\Omega''_{nm}) \tag{5.53}$$

Values of Q_{2m} corresponding to the Ω'_{2m} of Figure 5.11 are shown in Figure 5.12. In plotting Figure 5.12, we again took the first three terms in each series of Equations 5.49, $R = 0$, $4 \leq B \leq 5$, and the material coefficients relevant to PZT-5.

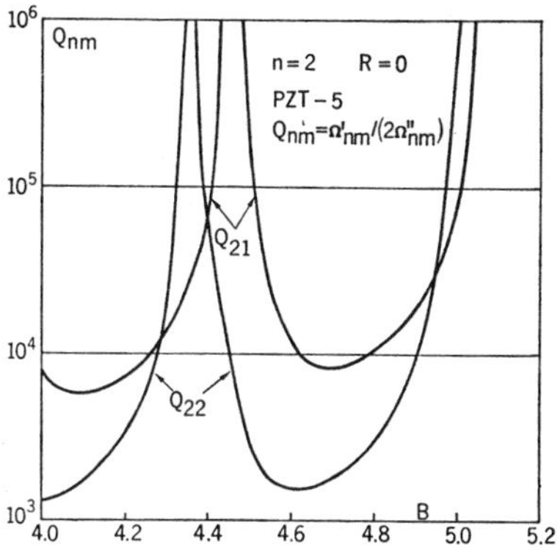

Figure 5.12. Quality factor of trapped-energy resonances for the second thickness-twist wave in a PZT-5 plate. Values were computed by coupling the first three thickness-twist wave solutions in the electroded and unelectroded regions.

From this figure, it is seen that Q_{21} and Q_{22} are nearly infinite for certain values of B and drop to the order of 10^3 or 10^4 in between those values. The Q's tend to infinity at electrode widths $2Bh$, which correspond to an integral number of $n = 1$ thickness-twist wavelengths. At such electrode widths, the energy conversion into the $n = 1$ wave at one edge interferes destructively with that at the other edge. This destructive interference prevents significant energy escape. On the other hand, Q minima occur where the electrode width is an odd multiple of $n = 1$ thickness-twist half-wavelengths. This condition leads to constructive interference of the $n = 1$ waves converted by each electrode edge and consequently yields maximum leakage. Mindlin and Lee have obtained Q versus B curves for $n = 1$ thickness-shear AT quartz plate trapped-energy resonators that resemble our Figure 5.12.[125] In the case of $n = 1$ thickness-shear quartz resonators, mode conversion into the flexure wave was found to be responsible for the energy leakage and subsequent noninfinite Q.

When Ω is complex, as in the case of the $n = 2$ thickness-twist trapped resonances, a curious thing happens to the first thickness-twist wavenumber Ξ_1' as given by Equations 5.36: It acquires a negative imaginary component,

$$\begin{aligned} \Xi_1' &= h^{-1}\pi(\Omega_{2m}'^2 - 2j\Omega_{2m}'\Omega_{2m}'' - \tfrac{1}{4})^{1/2} \\ &\doteq h^{-1}\pi\{(\Omega_{2m}'^2 - \tfrac{1}{4})^{1/2} - j\Omega_{2m}'\Omega_{2m}''(\Omega_{2m}'^2 - \tfrac{1}{4})^{-1/2}\} \end{aligned} \tag{5.54}$$

Upon substituting this wavenumber into the expression for $u_{3(1)}$ outside the electroded region (Equation 5.41), we find that $u_{3(1)}$ as a function of x_1 is a growing exponential! At first, we might think this behavior is a result of taking the wrong sign in a square root somewhere. However, reflection on the nature of our solution of the present problem can give us assurance that we have not actually made an error in arriving at this result.

The $n = 1$ wave which exponentially grows with x_1 is continually carrying energy out of the trapping region. However, this means the oscillations in the trapped region are effectively damped, and thus the rate of energy leak into the $n = 1$ wave is also damped. Consequently, if material losses are neglected, as has been implicitly done here so far, we should expect the $n = 1$ wave amplitude to grow as we recede from the trapping region with time held constant: In receding from the trapping region, we are looking at the leak energy which escaped earlier and earlier in time, when the trapping region was vibrating harder and harder.

In actual practice, we would probably never be able to detect energy leakage in $n = 2$ PZT-5 trapped-energy resonators, however, because the minimum Q associated with leakage in Figure 5.12 is about 10^3. On the other hand, the elastic material Q of PZT-5 is only about 50. In other words, energy loss because of material dissipation would always be at least 20 times as great as loss due to leakage in PZT-5 thickness-twist trapped-energy resonators.

We can achieve quite accurate solutions for the $n = 2$ trapped resonances simply by considering only the $n = 2$ and $n' = 2$ terms in Equations 5.49. This operation corresponds physically to ignoring mode conversion at the electrode edges and leads to a 2×2 determinant which is zero at the approximate Ω_{2m}. Figure 5.10 for the $n = 1$ trapped resonances was plotted using this same assumption of negligible mode conversion. Of course, Ω_{2m} computed in this approximation will lack its imaginary component, but this is not really significant for the reasons discussed in the previous paragraph. We have found

$(\Omega_{2m} - \Omega_{co})/(\Omega'_{co} - \Omega_{co})$ does not change by more than 1 percent in absolute value when one omits all but the $n = 2$ terms. Results for $(\Omega_{2m} - \Omega_{co})/(\Omega'_{co} - \Omega_{co})$ for PZT-5 with $R = 0$ and $0 \leq B \leq 10$ may be seen in Figure 5.13 as computed in this no-mode-conversion approximation.

Similar approximate values for $(\Omega_{nm} - \Omega_{co})/(\Omega'_{co} - \Omega_{co})$ may be determined for the higher order ($n > 2$) thickness-twist trapped-energy resonances. However, for $n > 1$, $(\Omega_{nm} - \Omega_{co})/(\Omega'_{co} - \Omega_{co})$ depends so weakly on n that the differences cannot be shown graphically. This means that while Figure 5.13 was plotted for the $n = 2$ case, it is an adequate representation of the trapped-resonance spectrum for any value of $n > 1$ in PZT-5 plates with $R = 0$ and $0 \leq B \leq 10$.

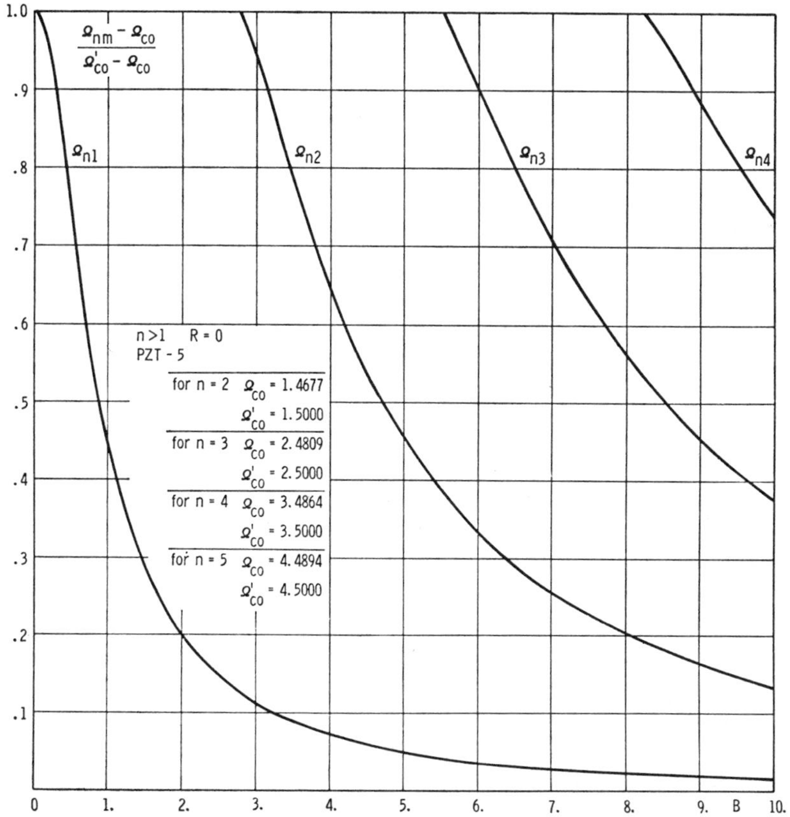

Figure 5.13. Trapped-energy resonance spectrum of the higher thickness-twist waves for a PZT-5 plate. Values were computed by coupling the nth thickness-twist wave solutions in the electroded and unelectroded regions.

5.2.5. Bechmann's Number

In the approximation that no energy is scattered out of the nth thickness-twist wave, Equations 5.49 become

$$\begin{aligned} &A_n I_{nn} \cos L_n B - A_n' = 0 \\ &A_n L_n \sin L_n B\left(-I_{nn} + J_{nn}[k^{(1)}]^2 \frac{\sin \mathscr{K}_n}{\sinh L_n}\right) - A_n' j L_n' = 0 \end{aligned} \tag{5.55}$$

These equations will have a nontrivial solution for A_n and A_n' only when the coefficient matrix has a vanishing determinant; this vanishing determinant yields the approximate Ω_{nm}. In the frequency band where Ω_{nm} can occur, L_n will be real and L_n' imaginary,

$$L_n' = j\pi\{(n - \tfrac{1}{2})^2 - \Omega^2\}^{1/2} = jl_n' \tag{5.56}$$

Consequently, the Ω_{nm} are determined in this approximation by Equation 5.56 and the secular equation associated with 5.55,

$$\tan L_n B = \frac{I_{nn} l_n'}{(I_{nn} - J_{nn}[k^{(1)}]^2 \sin \mathscr{K}_n / \sinh L_n) L_n} \tag{5.57}$$

In considering Equation 5.57, we should bear in mind that L_n, l_n', and $\mathscr{K}_n$ all depend on Ω. The number of roots Ω_{nm} which Equation 5.57 possesses for any given B may be fairly easily estimated. Usually the term in J_{nn} in the denominator of the right side of Equation 5.57 is smaller than I_{nn}. If we omit the term in J_{nn}, Equation 5.57 reduces to

$$\tan L_n B = \frac{l_n'}{L_n} \tag{5.58}*$$

We now sketch graphically $\tan L_n B$ and l_n'/L_n as functions of L_n so that the points of intersection of those two functions on that graph yield the resonant frequencies Ω_{nm}. This is done in Figure 5.14.

The function $\tan L_n B$ crosses the horizontal axis at $L_n = 0$, π/B, $2\pi/B$, etc. On the other hand, Equations 5.26 and 5.56 may be combined to express l_n'/L_n as

$$\frac{l_n'}{L_n} = \frac{1}{L_n}\{\pi^2(n - \tfrac{1}{2})^2 - (L_n^2 + \mathscr{K}_n^2)\}^{1/2} \tag{5.59}$$

When L_n is zero, Ω is the nth cutoff frequency in the electroded region $\Omega_{co(n)}$, and $\mathscr{K}_n$ is given by Equation 5.26 as $\pi\Omega_{co(n)}$. Moreover, numerical computations show that $\mathscr{K}_n$ does not vary appreciably in

* It is interesting to note that this is the classic energy-trapping resonance equation — see Equation 21 in Reference 111, Equation 12 in Reference 118, Equation 10 in Reference 117, or Equation 6 in Reference 122.

the band of interest $\Omega_{co(n)} < \Omega < (n - \frac{1}{2}) = \Omega'_{co(n)}$. We can thus approximate l'_n/L_n as

$$\frac{l'_n}{L_n} = \frac{1}{L_n}\left\{\pi^2[(n - \tfrac{1}{2})^2 - \Omega^2_{co(n)}] - L_n^2\right\}^{1/2} \tag{5.60}$$

As a function of L_n, this formula resembles a hyperbola (see Figure 5.14). For small L_n, the l'_n/L_n approaches infinity, and l'_n/L_n decreases monotonically with L_n as L_n increases. However, for some finite L_n, say $(L_n)_{\max}$, the l'_n/L_n becomes zero, and for larger L_n, the l'_n/L_n is imaginary.

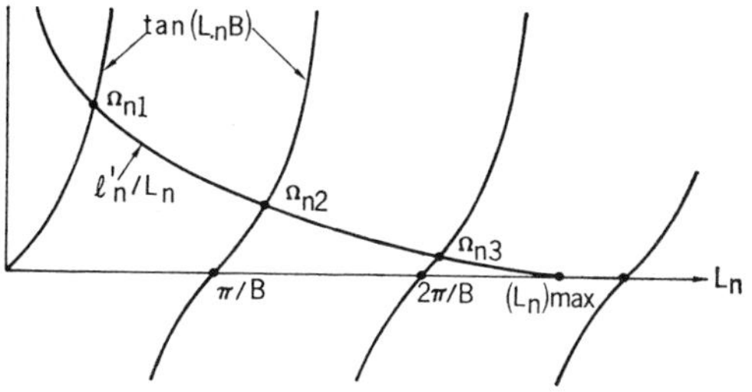

Figure 5.14. Plots of tan L_nB and l'_n/L_n versus L_n. The intersections of the two functions determine the approximate thickness-twist trapped-energy resonances.

Consequently, beyond this $(L_n)_{\max}$, l'_n/L_n cannot intersect tan L_nB, which remains pure real. An expression for $(L_n)_{\max}$ may be obtained from Equation 5.60;

$$(L_n)_{\max} = \pi(\Omega'^2_{co(n)} - \Omega^2_{co(n)})^{1/2} \tag{5.61}$$

Based on the graphical construction in Figure 5.14, we see that Equation 5.58 will thus have $r + 1$ solutions, where r is the greatest integer such that

$$(L_n)_{\max} > r\pi/B \tag{5.62}$$

This inequality may be rewritten with the use of Equation 5.61,

$$r < B\sqrt{2}\,\Omega'_{co(n)}\Delta_n = B\sqrt{2}\,(n - \tfrac{1}{2})\Delta_n \tag{5.63}$$

where

$$\Delta_n = \sqrt{\frac{\Omega'_{co(n)} - \Omega_{co(n)}}{\Omega'_{co(n)}}} \tag{5.64}$$

In order to avoid spurious responses, the electrode width $2Bh$ should be small enough that Equation 5.58 has only one solution. The inequality of 5.63, indicates that this requirement will be met if

$$B < \frac{1}{\sqrt{2}\,(n - \frac{1}{2})} \cdot \frac{1}{\Delta_n} \tag{5.65}$$

The coefficient of Δ_n^{-1} in equations such as Equation 5.65 is usually referred to as Bechmann's number for the geometry and material in question. Thus, Bechmann's number for trapping just one thickness-twist resonance of order n in a ferroelectric ceramic plate is $[\sqrt{2}\,(n - \frac{1}{2})]^{-1}$.

5.2.6. Normalized Dynamic Capacitances

Let us now consider the equivalent electrical circuit of the electrode strips shown in Figure 5.3. If these strips are actually infinite in the x_3-direction, their admittance also must be infinite. However, this difficulty may be overcome by evaluating only the admittance associated with a finite segment of the electrode strips. We shall do this here and for convenience shall let the x_3-length of the segment in question be one unit.

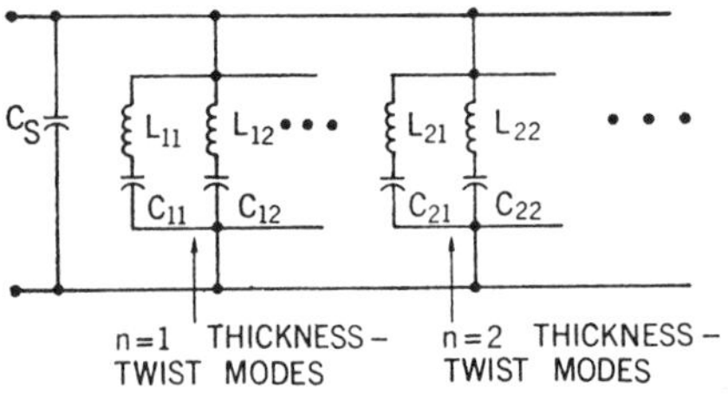

Figure 5.15. Equivalent circuit for a unit length segment of the electrode strips shown in Figure 5.3.

Under these assumptions, the finite electrode strip may be represented electrically over all frequencies by the equivalent circuit shown in Figure 5.15. Note that this circuit is merely an adoption of that shown in Figure 4.11 — the series LC resonators have been grouped in the present circuit to emphasize that the trapped-energy resonances occur in bands. One band of resonances is associated with each thickness-twist wave.

The present equivalent circuit is to have the nmth resonant branch correspond to Ω_{nm}: in view of Equation 5.22, this means

$$\frac{1}{L_{nm}C_{nm}} = \Omega_{nm}^2 \frac{\pi^2 c_{44}^D}{\rho h^2} = \omega_{nm}^2 \tag{5.66}$$

The admittance of the equivalent circuit in Figure 5.15 may then be written as

$$Y = j\omega C_S + j\omega \sum_{n,m} \frac{C_{nm}\omega_{nm}^2}{\omega_{nm}^2 - \omega^2} \tag{5.67}$$

This expression for Y is to equal the admittance given by Equation 4.40, which also applies here in the energy-trapping case,

$$Y = j\omega C_S + j\omega \sum_{n,m} \frac{(Q^{(nm)})^2}{\omega_{nm}^2 - \omega^2} \tag{5.68}*$$

In Equations 5.67 and 5.68, as in Figure 5.15, the C_S is the clamped or high-frequency capacitance of the unit electrode segments

$$C_S = \epsilon_{11}^S b/h = \epsilon_{11}^S B \tag{5.69}$$

Additionally, $Q^{(nm)}$ is the charge on the top electrode segment due to the *nm*th short-circuit trapped resonance: In the no-mode-conversion approximation, this becomes

$$Q^{(nm)} = -\int_{-b}^{b} \left[e_{24} \frac{\partial u_{3(n)}}{\partial x_2} - \epsilon_{11}^S \frac{\partial \phi_{(n)}}{\partial x_2} \right]_{\substack{\omega=\omega_{nm} \\ x_2=h}} dx_1 \tag{5.70}$$

If $u_{3(n)}$ and $\phi_{(n)}$ are substituted from Equations 5.28 and 5.29, this expression for $Q^{(nm)}$ becomes

$$\begin{aligned} Q^{(nm)} &= 2A_n e_{24} \coth \Xi_n h \sin \Xi_n b \sin K_n h\big|_{\omega=\omega_{nm}} \\ &= 2A_n e_{24} \coth L_n \sin L_n B \sin \mathscr{K}_n\big|_{\Omega=\Omega_{mn}} \end{aligned} \tag{5.71}$$

The quantity A_n in Equation 5.71 may be determined by applying the normalization condition, Equations 4.30 and 4.32:

$$1 = \int_{-\infty}^{\infty} dx_1 \int_{-h}^{h} dx_2 \rho (u_{3(n)})^2 \bigg|_{\omega=\omega_{nm}} \tag{5.72}$$

Insertion of $u_{3(n)}$ from Equations 5.28 and 5.41 and elimination of A_n' by means of the first of Equation 5.55 then yields

$$\begin{aligned} A_n\big|_{\Omega=\Omega_{nm}} = (\rho b h)^{-1/2} \bigg\{ &\left[1 - \frac{\sin 2\mathscr{K}_n}{2\mathscr{K}_n}\right]\left[1 + \frac{\sin 2L_n B}{2L_n B}\right] \\ &+ I_{nn}^2 \frac{\cos^2 L_n B}{l_n' B} \bigg\}^{-1/2} \bigg|_{\Omega=\Omega_{nm}} \end{aligned} \tag{5.73}$$

The piezoelectric strength of resonance *nm* is proportional to the dynamic capacitance of that resonance C_{nm}. This capacitance may be

* Subscripts are not used on Y and $Q^{(nm)}$ in this section because the device in question possesses just two electrodes. We shall consider the bottom electrode to be the reference electrode 0, and the top electrode to be electrode 1. Thus $Q^{(nm)}$ actually means $Q_1^{(nm)}$ and Y means Y_{11}.

obtained from Equations 5.67 and 5.68 as

$$C_{nm} = (Q^{(nm)})^2/\omega_{nm}^2 \tag{5.74}$$

A high C_{nm} indicates that the admittance level of the electrode strips in the vicinity of ω_{nm} is high

However, it is inconvenient to present numerical data on C_{nm} directly because it is not a dimensionless quantity and depends on such things as ρ, b, h, e_{24}, etc Numerical data on the strength of resonance *nm* is better given as the normalized dynamic capacitance. We shall define this dimensionless quantity as

$$C_{nmN} = \frac{C_{nm}}{(\epsilon_{11}^T - \epsilon_{11}^S)b/h} = \frac{C_{nm}}{\epsilon_{11}^T[k^{(1)}]^2} \cdot \frac{1}{B} \tag{5.75}$$

Once values for C_{mnN} are given, it is immediately possible to determine

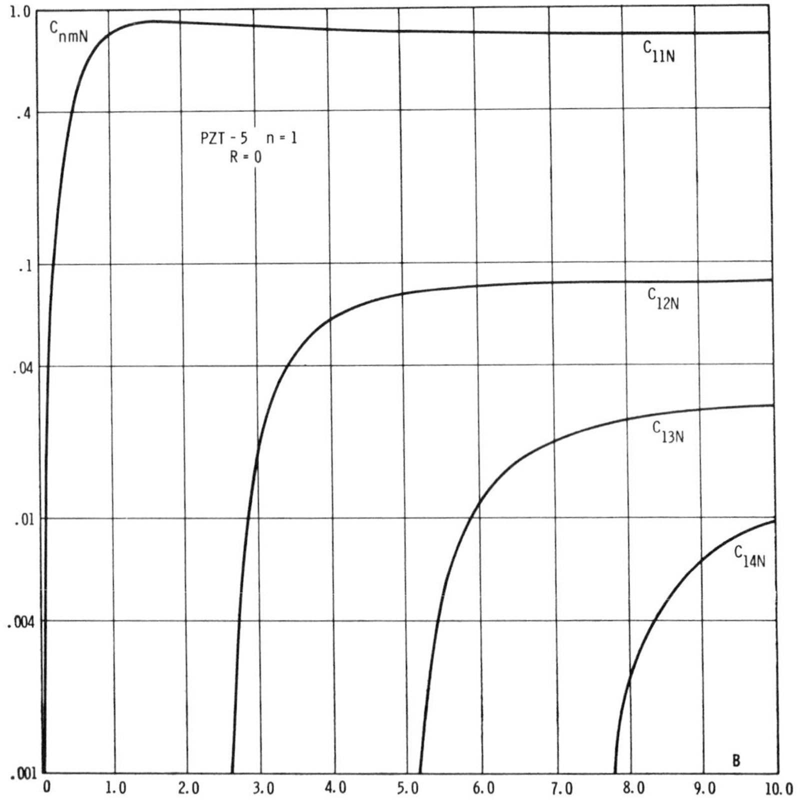

Figure 5.16. Normalized dynamic capacitances for trapped resonances of the first thickness-twist wave in a PZT-5 plate.

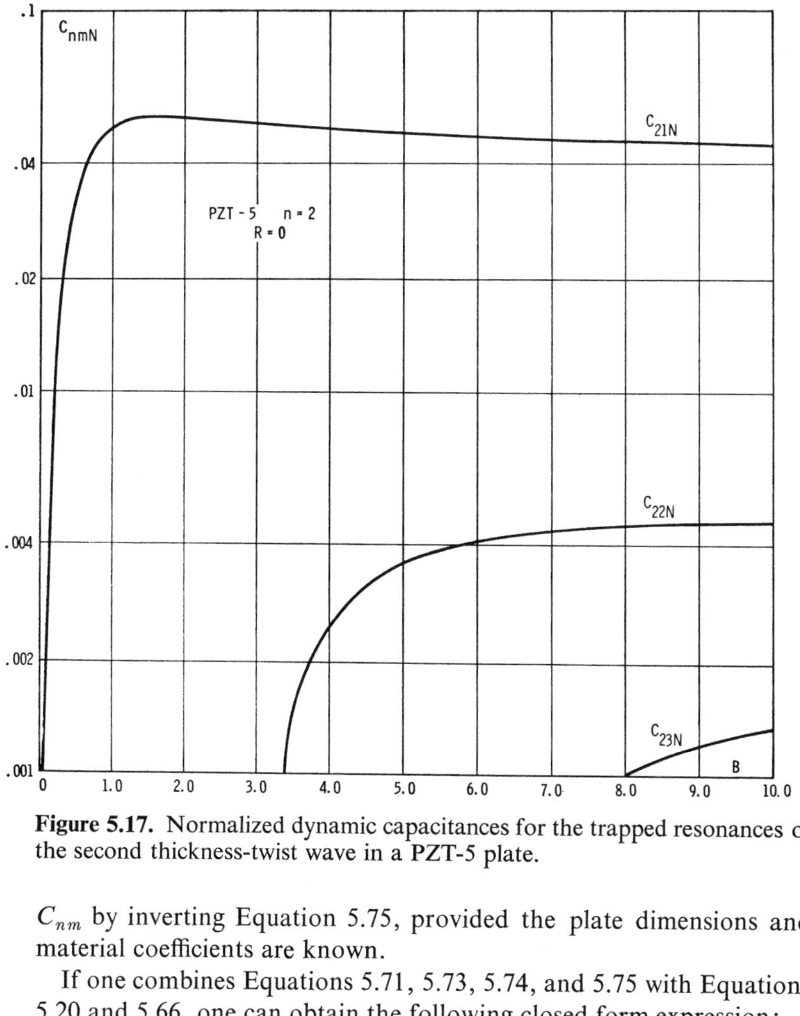

Figure 5.17. Normalized dynamic capacitances for the trapped resonances of the second thickness-twist wave in a PZT-5 plate.

C_{nm} by inverting Equation 5.75, provided the plate dimensions and material coefficients are known.

If one combines Equations 5.71, 5.73, 5.74, and 5.75 with Equations 5.20 and 5.66, one can obtain the following closed form expression:

$$C_{nmN} = \left. \frac{4(1-[k^{(1)}]^2)\coth^2 L_n \sin^2 L_n B \sin^2 \mathscr{K}_n}{B^2\pi^2\Omega_{nm}^2\left\{\left[1-\frac{\sin 2\mathscr{K}_n}{2\mathscr{K}_n}\right]\left[1+\frac{\sin 2L_nB}{2L_nB}\right]+I_{nn}^2\frac{\cos^2 L_nB}{l'_nB}\right\}}\right|_{\Omega=\Omega_{nm}} \tag{5.76}$$

Values for C_{nmN} for $n = 1, 2, 3$ are shown in Figures 5.16 to 5.18, respectively. These figures are based on a PZT-5 plate with $R = 0$ and $0 \leq B \leq 10$.

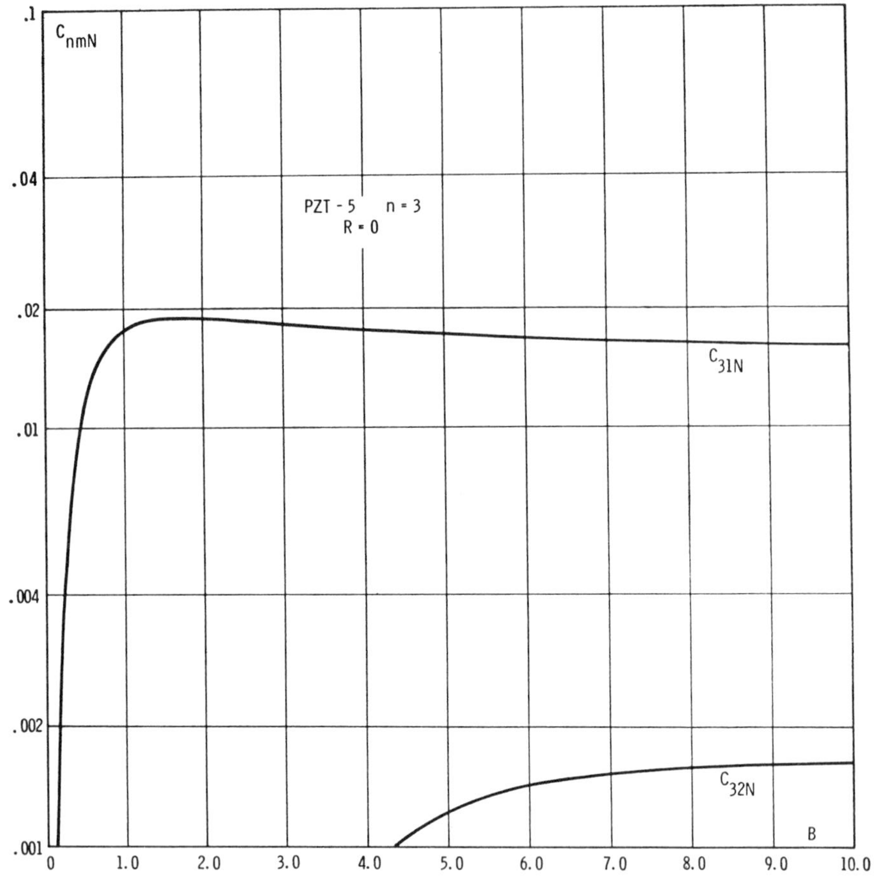

Figure 5.18. Normalized dynamic capacitances for the trapped resonances of the third thickness-twist wave in a PZT-5 plate.

In concluding this section, we point out that a derivation analogous to that establishing Equation 4.103 will show

$$\lim_{B \to \infty} \sum_{n,m} C_{nmN} = 1 \tag{5.77}$$

5.3. Thickness-Dilation Modes

In the previous section, we considered energy trapping in a ferroelectric ceramic plate polarized in its own plane. There the types of waves that become trapped are thickness-twist and thickness-shear waves.

However, it is actually rather difficult to polarize a large ferroelectric ceramic plate in its own plane. Also, the idealized one-dimensionality of our foregoing analysis is not easy to realize experimentally.

Figure 5.19 illustrates a somewhat more practical ferroelectric ceramic energy trapping configuration. Here a finite electrode pair is located on the top and bottom surfaces of a large, but finite, plate. The

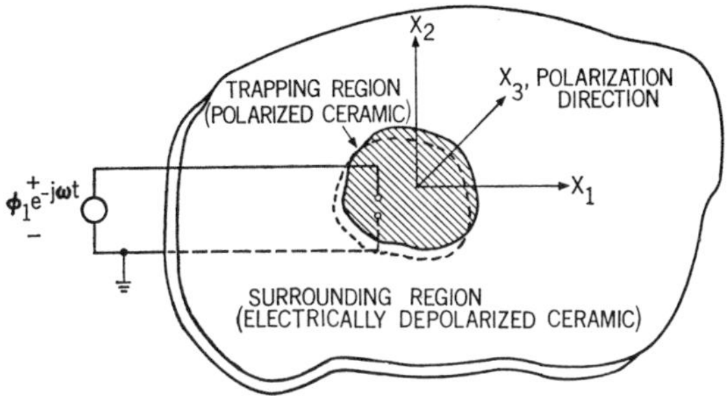

Figure 5.19. A configuration for trapping thickness-dilational waves in a finite ferroelectric ceramic plate.

entire plate is assumed to be originally in an electrically depoled condition.* Application of a large voltage across the electrode pair may then be used to polarize the electroded region in the x_3 or thickness direction. This operation will leave the surrounding ceramic region unaffected.

Once the ceramic in the electroded region has been polarized, one may drive the plate piezoelectrically by application of a small rf voltage to the electrodes. If this is done at the proper frequency, one

* This condition is achieved as follows: First, electrode the entire top and bottom surfaces of the plate. Second, apply a large voltage across the electrodes until a maximum polarization in the $+x_3$-direction is reached. Third, reverse the large voltage until the polarization in the plate has switched exactly halfway from the $+x_3$- to the $-x_3$-direction and then turn off the voltage. Last, remove the electrodes from the portion of the plate specified in Figure 5.19 to be depoled. An electrically depoled plate, as with a virgin or thermally depoled plate, is piezoinactive. However, an electrically depoled plate retains an axis of elastic anisotropy along x_3. Also, it is possible to switch the polarization differentially in two parts of an electrically depoled plate without cracking the plate. Differential switching, on the other hand, always cracks thermally depoled or virgin samples.

will excite thickness-dilational waves in the plate. Normally, the thickness-dilational waves have a higher cutoff frequency in the surrounding region than in the electroded region. Consequently, dilation-wave trapped-energy resonances associated with the electroded region will be found to occur between the two cutoff frequencies.

The advantage of this structure over that shown in Figure 5.3 is that here it is quite simple to give this sample the required polarization: That is not the case in Figure 5.3. However, dilation waves possess vastly more complicated motions than thickness-twist or thickness-shear waves. In other words, while it is very simple to achieve dilation-wave energy trapping experimentally, it is very difficult to understand this phenomenon quantitatively.

The mathematics involved in this topic, in fact, is so tedious that we will omit complete derivations in the present section. Instead, we will merely present the basic assumptions and approximations to be used and will then write the wave equations to which these assumptions and approximations lead. This information will be presented in sufficient detail that the reader should have no difficulty in applying it, however.

In conclusion, we will demonstrate a technique for solving these equations for dilation waves by using as an example the structure shown in Figure 5.19 in the special case of a circular electrode region.

5.3.1. The Wave Equations in the Electroded Region

Mindlin and Medick have developed an approximation procedure for treating thickness-dilational waves in isotropic plates.[126] The results which we are about to state without proof are essentially an extension of that earlier work to thickness-poled ferroelectric plates. Anyone seriously interested in reproducing the derivation of our equations is advised first to go through the rather complete derivation found in Mindlin and Medick's paper. It should then be fairly obvious how we have generalized this derivation to include anisotropy and piezoelectric effects.

In particular, let us first consider the electroded region of the plate shown in Figure 5.19. We shall assume that this region has been polarized in the x_3 or thickness direction and that the top electrode now is being subjected to a small-signal rf voltage $\phi_1 e^{-j\omega t}$ while the bottom electrode is held at ground (see Figure 5.19). We shall consider the center of the plate to be the x_1x_2-plane, the electrodes to be negligibly thin, the surfaces of the plate to be mechanically free, and the surfaces of the plate to be located at $x_3 = \pm h$.

The particle displacement and electric potential may be expanded across the thickness dimension in Legendre polynomials* as

$$
\begin{aligned}
u_j &= \sum_{\alpha=0}^{\infty} U_j^{(\alpha)}(x_1,x_2)P_\alpha(x_3/h) \\
\phi &= \sum_{\alpha=0}^{\infty} \Phi^{(\alpha)}(x_1,x_2)P_\alpha(x_3/h)
\end{aligned}
\tag{5.78}
$$

Equations 5.78 and subsequent formulas are understood to contain an $e^{-j\omega t}$ dependence which is suppressed. This expansion corresponds to Equation 1 of Mindlin and Medick. Note that $U_j^{(\alpha)}$ and $\Phi^{(\alpha)}$ are functions of x_1 and x_2 only.

The reason for expanding u_j and ϕ as in Equation 5.78 is to permit elimination of x_3-dependence from the problem. In other words, it is desired to replace the four exact three-dimensional partial differential equations of piezoelectric motion (see Equations 4.3 and 4.4) with more than four approximate two-dimensional partial differential equations. Whereas the variables in the exact differential equations are $u_j(x_1,x_2,x_3)$ and $\phi(x_1,x_2,x_3)$, Equations 5.78 permit replacement of these variables with $U_j^{(\alpha)}(x_1,x_2)$ and $\Phi^{(\alpha)}(x_1,x_2)$. In this way, the desired elimination of x_3 from the equations of motion is effected.

Dilation waves in this formulation are characterized by even α in $U_1^{(\alpha)}$ and $U_2^{(\alpha)}$ and by odd α in $U_3^{(\alpha)}$ and $\Phi^{(\alpha)}$. If we bear this in mind, the two-dimensional wave equations for the electroded region may be derived from Equation 5.78 upon step by step insertion of piezoelectric and anisotropic effects in Equations 1 to 35 of Mindlin and Medick. Results are

$$
\begin{aligned}
&\tfrac{1}{2}(c_{11}^E + c_{12}^E)U_{i,ij}^{(0)} + c_{66}^E\nabla^2 U_j^{(0)} + \frac{K_1 c_{13}^E U_{3,j}^{(1)}}{h} + \rho\omega^2 U_j^{(0)} \\
&\qquad = -\frac{(e_{31}\Phi^{(1)})_{,j}}{h} \\
&K_2^2 c_{55}^E \nabla^2 U_3^{(1)} - \frac{3c_{13}^E K_1 U_{i,i}^{(0)}}{h} - \frac{3K_1^2 c_{33}^E U_3^{(1)}}{h^2} + \frac{3c_{55}^E K_2^2 U_{i,i}^{(2)}}{h} + \rho K_3^2 \omega^2 U_3^{(1)} \\
&\qquad = -K_2\nabla^2(e_{15}\Phi^{(1)}) + \frac{3e_{33}K_1K_5\Phi^{(1)}}{h^2} \\
&c_{66}^E\nabla^2 U_j^{(2)} + \frac{E'(1+\nu^D)}{2}U_{i,ij}^{(2)} \\
&\qquad - \frac{5}{h}[K_2^2 c_{55}^E(U_{3,j}^{(1)} + 3h^{-1}U_j^{(2)})] + \rho K_4^2\omega^2 U_j^{(2)} \\
&\qquad = \frac{5}{h}K_2(e_{15}\Phi^{(1)})_{,j}
\end{aligned}
\tag{5.79}
$$

where $i, j = 1$ and 2.

* It is also possible to expand u_j and ϕ in power series of (x_3/h).[127] This superficially simpler approach, however, ultimately leads to more complicated results because the powers of (x_3/h) lack the orthogonality inherent in the $P_\alpha(x_3/h)$.

In Equations 5.79, we have defined

$$\nabla^2 = \frac{\partial^2}{\partial x_1^2} + \frac{\partial^2}{\partial x_2^2} \tag{5.80}$$

$$E' = c_{11}^D - c_{13}^{D^2}/c_{33}^D \tag{5.81}$$

and

$$\nu^D = -s_{12}^D/s_{11}^D \tag{5.82}$$

Additionally in Equations 5.79, the series for $U_1^{(\alpha)}$ and $U_2^{(\alpha)}$ have been truncated after $\alpha = 2$, while the series for $U_3^{(\alpha)}$ and $\Phi^{(\alpha)}$ have been truncated after $\alpha = 1$. The applied rf voltage ϕ_1 must then be related to $\Phi^{(1)}$ by $\Phi^{(1)} = \phi_1/2$, as one may easily demonstrate. Truncating the series in this manner will obviously lead to discrepancies between the exact solution and the solution obtained from Equations 5.79. The quantities K_1, K_2, K_3, K_4, and K_5 are factors on the order of unity which have been arbitrarily introduced into Equations 5.79. These factors will be adjusted later on so as to minimize the error in the solution of Equations 5.79 resulting from series truncations.

In the three equations of 5.79, expressions in $\Phi^{(1)} = \phi_1/2$ appearing on the right represent inhomogeneous or driving terms. It is interesting to compare these terms with those occurring analogously in the Green's function analysis of Chapter 2. For instance, the first equation of 5.79 is primarily a wave equation for $U_j^{(0)}$. This is the planar motional component which is not associated with any thickness dependence. In other words, $U_j^{(0)}$ may be regarded as describing the contour extensional motion of the sample in Figure 5.19. The driving term in this first equation of 5.79 is $-(e_{31}\Phi^{(1)}/h)_{,j}$. This expression bears an obvious resemblance to the electrical driving term in Equation 2.51, the planar wave equation for contour extensional motion in a plate of arbitrary piezoelectrical material: That driving term was $(\tilde{e}_{3ij}E_3)_{,i}$. For example, both $(e_{31}\Phi^{(1)}/h)_{,j}$ and $(e_{3ij}E_3)_{,i}$ represent singularity functions of force concentration acting only at the electrode edge.

Similarly, the driving term in e_{33} in the second of Equations 5.79 is identifiable with the thickness dilational forcing term of Equation 2.135. For both Equations 5.79 and 2.135, these forcing terms are localized at the x_3 ceramic surfaces.

The driving terms containing e_{15} in Equations 5.79 do not have counterparts in the Green's function discussion. However, even these terms are singularity functions acting, as with the term in e_{31}, around the electrode edge exclusively.

5.3.2. Determination of the K_1 to K_4

Values are assigned the arbitrary factors K_1 to K_4 of Equations 5.79 as follows: First solve both Equations 5.79 and the exact equations of motion for the electroded region in Figure 5.19 in the special case of a straight-crested wave with the two electrodes grounded and shorted together. Then adjust K_1 to K_4 so that the solution of Equations 5.79 coincides with the exact solution in as many ways as possible.

In the case of a straight-crested wave propagating along x_1, we have $\partial/\partial x_2 = 0$, and thus the approximate equations of motion, Equations 5.79, become

$$\begin{aligned}
&c_{11}^E U_{1,11}^{(0)} + \frac{K_1 c_{13}^E}{h} U_{3,1}^{(1)} + \rho\omega^2 U_1^{(0)} = 0 \\
&c_{66}^E U_{2,11}^{(0)} + \rho\omega^2 U_2^{(0)} = 0 \\
&K_2^2 c_{55}^E U_{3,11}^{(1)} - \frac{3c_{13}^E K_1}{h} U_{1,1}^{(0)} - \frac{3K_1^2 c_{33}^E}{h^2} U_3^{(1)} \\
&\qquad + \frac{3c_{55}^E K_2^2}{h} U_{1,1}^{(2)} + \rho K_3^2\omega^2 U_3^{(1)} = 0 \\
&E' U_{1,11}^{(2)} - \frac{5}{h}[K_2^2 c_{55}^E (U_{3,1}^{(1)} + 3h^{-1} U_1^{(2)})] + \rho K_4^2 \omega^2 U_1^{(2)} = 0^* \\
&c_{66}^E U_{2,11}^{(2)} - \frac{15}{h^2} U_2^{(2)} + \rho K_4^2 \omega^2 U_2^{(2)} = 0
\end{aligned} \tag{5.83}$$

The quantity $\Phi^{(1)} = \phi_1/2$ has here been set equal to zero, as we have specified that K_1 to K_4 are to be determined under short-circuit conditions.

Note that Equation 5.83 consists of three coupled equations in $U_1^{(0)}$, $U_3^{(1)}$, and $U_1^{(2)}$; and of two uncoupled equations in $U_2^{(0)}$ and $U_2^{(2)}$. We shall be primarily concerned with the coupled equations, as they characterize dilation waves. The two decoupled equations represent uniform shear and thickness-twist waves,† which are not of interest to us here as they are not piezoactive.

Now let us assume that the straight-crested approximate solutions for $U_1^{(0)}$, $U_3^{(1)}$, and $U_1^{(2)}$ may be written as

$$U_1^{(0)} = A \sin \Xi x_1 \qquad U_3^{(1)} = B \cos \Xi x_1 \qquad U_1^{(2)} = C \sin \Xi x_1 \tag{5.84}$$

* For most ferroelectric ceramics it is true that $E'(1 - \nu^D)/2 \doteq c_{66}^E$, ± 2 percent.

† Thickness-twist wave solutions resemble thickness-shear solutions in that both have particle displacement and propagation in the plane of the plate. However, in thickness-twist solutions, displacement and propagation are perpendicular, whereas they are parallel in thickness-shear solutions (see Figure 11 of Reference 111).

If these expressions are substituted into the three coupled equations of 5.83, we find

$$\begin{aligned}
&\left(z^2 \frac{c_{11}^E}{c_{55}^E} - \Omega^2\right) A + \frac{2K_1 c_{13}^E z}{\pi c_{55}^E} B = 0 \\
&\frac{2K_1 c_{13}^E z}{\pi c_{55}^E} A + \left(\frac{z^2 K_2^2}{3} + \frac{4K_1^2 c_{33}^E}{\pi^2 c_{55}^E} - \frac{\Omega^2 K_3^2}{3}\right) B - \frac{2K_2^2 z}{\pi} C = 0 \qquad (5.85) \\
&-\frac{2K_2^2 z}{\pi} B + \left(\frac{12K_2^2}{\pi^2} + \frac{E' z^2}{5c_{55}^E} - \frac{\Omega^2 K_4^2}{5}\right) C = 0
\end{aligned}$$

where z is a dimensionlessly normalized wavenumber

$$z = 2\Xi h/\pi \qquad (5.86)$$

and Ω is a dimensionlessly normalized frequency

$$\Omega = \frac{2h\omega}{\pi} \sqrt{\frac{\rho}{c_{55}^E}} \qquad (5.87)$$

For convenience, we can also write Equations 5.85 as

$$\begin{aligned}
a_{11}A + a_{12}B \qquad\qquad &= 0 \\
a_{12}A + a_{22}B + a_{23}C &= 0 \qquad (5.88) \\
a_{23}B + a_{33}C &= 0
\end{aligned}$$

Equations 5.88 possess nontrivial solutions for A, B, and C only when the associated coefficient matrix has a vanishing determinant. For each value of Ω, one will find three values of z^2 that meet this condition. We shall represent these values as $z_{(m)}(\Omega)$ where $m = 1, 2, 3$. In this way, Equations 5.88 yield approximate curves for the first three branches of the dispersion relation for straight-crested dilation waves in a short-circuited ferroelectric ceramic plate.

The first set of conditions to be applied in the evaluation of K_1 to K_4 is that the three approximate dispersion branches defined by Equation 5.88 should have cutoff frequencies coinciding with the exact cutoff frequencies. Cutoff frequencies will occur whenever $z = 0$. Substitution of this condition into Equations 5.85 and 5.88 yields

$$\begin{vmatrix}
-\Omega^2 & 0 & 0 \\
0 & \dfrac{4K_1^2 c_{33}^E}{\pi^2 c_{55}^E} - \dfrac{\Omega^2 K_3^2}{3} & 0 \\
0 & 0 & \dfrac{12K_2^2}{\pi^2} - \dfrac{\Omega^2 K_4^2}{5}
\end{vmatrix} = 0 \qquad (5.89)$$

Roots of Equation 5.89 occur at

$$\Omega^2_{co(m)} = 0, \frac{12K_1^2c_{33}^E}{\pi^2K_3^2c_{55}^E}, \frac{60K_2^2}{\pi^2K_4^2} \tag{5.90}$$

with corresponding solutions for $A:B:C$ being, respectively,

$$A:B:C = 1:0:0, \quad 0:1:0, \quad 0:0:1 \tag{5.91}$$

Reference to Equations 5.78 and 5.84 indicates that the first of these solutions corresponds to the planar dilational or contour-extensional type wave. This has cutoff at $\Omega^2 = 0$ in the exact solution as well as in the present approximate analysis. Consequently, the cutoff of the $z_{(1)}$-wave does not provide us with any information for determining K_1 to K_4.

The second solution in Equations 5.90 and 5.91 approximates the cutoff of the first thickness-dilational wave — the wave we are primarily interested in here. This cutoff frequency should coincide exactly with the thickness-dilational resonance of an infinite, short-circuited ferroelectric ceramic plate. That resonance, when found exactly, has the frequency

$$\Omega^2_{co(2)} = \frac{4\kappa^2c_{33}^D}{\pi^2c_{55}^E} \tag{5.92}$$

where κ is the first root of

$$\tan\kappa = \kappa/k_t^2 \tag{5.93}$$

and where k_t^2 is the thickness-dilational coupling factor, $h_{33}^2/(c_{33}^D\beta_{33}^S)$. Comparison of $\Omega_{co(2)}$ as given in Equations 5.90 and 5.92 then provides us with the first condition on the K_1 to K_4,

$$\frac{K_1}{K_3} = \kappa\sqrt{\frac{c_{33}^D}{3c_{33}^E}} \tag{5.94}$$

The third solution in Equations 5.90 and 5.91 represents an approximation to the cutoff of the first symmetric thickness-shear wave. This cutoff should agree with the exact solution for the first symmetric thickness-shear resonance of an infinite ferroelectric ceramic plate,

$$\Omega^2_{co(3)} = 4 \tag{5.95}$$

(This shear resonance is not piezoactive, and thus the short-circuit electrical boundary condition does not enter Equation 5.95 in any way.) Equating $\Omega_{co(3)}$ from 5.90 and 5.95, we have a second condition on

K_1 to K_4,

$$\frac{K_2}{K_4} = \frac{\pi}{\sqrt{15}} \tag{5.96}$$

Equations 5.94 and 5.96 give us two relations for the K_1 to K_4; we need two more before we can determine the K_1 to K_4 uniquely. In order to obtain these, let us match the approximate and exact derivatives of Ω with respect to $z_{(m)}$ at $\Omega_{co(m)}$.

For example, if we expand the determinant associated with Equation 5.88 and evaluate the derivative of Ω with respect to $z_{(1)}$ at $z_{(1)} = 0$, $\Omega = \Omega_{co(1)} = 0$, we find

$$\left.\frac{d\Omega}{dz_{(1)}}\right|_{\Omega=\Omega_{co(1)}} = N_1 = \sqrt{\frac{c_{11}^E}{c_{55}^E} - \frac{(c_{13}^E)^2}{c_{33}^E c_{55}^E}} \tag{5.97}$$

At $z_{(2)} = 0$, $\Omega = \Omega_{co(2)}$ and at $z_{(3)} = 0$, $\Omega = \Omega_{co(3)}$, the first derivative of Ω with respect to $z_{(m)}$ as computed from Equation 5.88 is zero. (This agrees identically with the exact solution.) However, the second derivative is finite in both cases,

$$\left.\frac{d^2\Omega}{dz_{(2)}^2}\right|_{\Omega=\Omega_{co(2)}} = N_2 = \frac{3}{K_3^2\Omega_{co(2)}^3}\left[\frac{4K_1^2c_{13}^{E^2}}{\pi^2 c_{55}^E} - \frac{\Omega_{co(2)}^4 K_2^2}{3(4-\Omega_{co(2)}^2)}\right] \tag{5.98}$$

and

$$\left.\frac{d^2\Omega}{dz_{(3)}^2}\right|_{\Omega=\Omega_{co(3)}} = N_3 = \frac{1}{2}\left[\frac{E'}{K_4^2 c_{55}^E} - \frac{K_2^2}{3K_1^2 c_{33}^E/(\pi^2 c_{55}^E) - K_3^2}\right] \tag{5.99}$$

It is possible to solve the corresponding exact equations numerically for $z_{(m)}(\Omega)$ in the case of straight-crested dilational waves in short-circuited ferroelectric ceramic plates.[128] (Closed form exact expressions, however, are not accessible.) From these numerical solutions, we can obtain exact values for N_1, N_2, and N_3. When this is done, it develops that the exact value for N_1 does not differ from Equation 5.97, and consequently Equation 5.97 provides no information on K_1 to K_4. Hence, we are left with Equations 5.94, 5.96, 5.98, and 5.99 which give four conditions for unique determination of K_1 to K_4.

The procedure given in Reference 128 for obtaining N_2 and N_3 exactly is admittedly rather tedious. However, if we neglect anisotropy, a fairly accurate value for N_2 and N_3 may be obtained by an alternative method which is much simpler.[129] In particular, if we define

$$k^2 = c_{33}^E/c_{55}^E \tag{5.100}$$

then N_2 and N_3 are approximated by

$$N_2 \doteq \frac{4k}{\pi}\left(\frac{\pi}{4} + \frac{4}{k^3}\cot\frac{\pi k}{2}\right)$$
$$N_3 \doteq \frac{1}{\pi}\left(\frac{\pi}{2} - \frac{4}{k}\tan\frac{\pi}{k}\right) \tag{5.101}$$

For PZT-5A,* N_2 and N_3 are given in Table 5.1, first, as computed exactly from the procedure described in Reference 128 and second, as

Table 5.1. Values for N_2, N_3, and K_1 to K_4 Based on PZT-5A*

	Computed from Reference 128	Computed from Equations 5.100 and 5.101
N_2	4.989	4.907
N_3	−2.579	−2.957
K_1	0.846	1.247
K_2	1.283	1.870
K_3	0.919	1.355
K_4	1.581	2.305

* Based on $c_{11}^E = 13.56 \times 10^{10}$, $c_{12}^E = 9.05 \times 10^{10}$, $c_{13}^E = 8.47 \times 10^{10}$, $c_{33}^E = 11.96 \times 10^{10}$, $c_{55}^E = 2.40 \times 10^{10}$, $c_{66}^E = 2.26 \times 10^{10}$, $e_{31} = -6.36$, $e_{33} = 13.20$, $e_{15} = 10.10$, $\epsilon_{33}^S = 887\epsilon_0$, $\epsilon_{11}^S = 840\epsilon_0$, and $\rho = 7750$ in MKS units. See Reference 130.

computed approximately from Equations 5.100 and 5.101. Also appearing in Table 5.1 are the two sets of values for K_1 to K_4 based on the two pairs of values for N_2 and N_3.

5.3.3. Determination of K_5

The factors K_1 to K_4 which we have just finished determining are introduced into Equations 5.79 to compensate for truncating the series in $U_i^{(\alpha)}$ after $\alpha = 1$ or 2. However, K_5 is introduced to compensate for truncating the series in $\Phi^{(\alpha)}$ after $\alpha = 1$. When the electrodes are shorted, $\Phi^{(1)} = \phi_1/2$ is zero in Equations 5.79, and K_5 drops out of the equations. Consequently, we must either use electrical boundary conditions other than a short circuit or we must apply some equation of motion other than those already given in Equations 5.79 to evaluate K_5.

Equations 5.79 are derived from Equations 5.78 and from the conservation of momentum, Equation 4.3. Use has not yet actually been

* Clevite Corporation, trade name.

made of the conservation of charge, Equation 4.4. Upon substitution of Equations 5.78 and repetition of steps (1) to (35) in the paper of Mindlin and Medick, conservation of charge becomes

$$K_2 e_{15}\nabla^2 U_3^{(1)} - \frac{3e_{31}}{h} U_{i,i}^{(0)} - \frac{3K_1K_5e_{33}}{h^2} U_3^{(1)} + \frac{3e_{15}K_2U_{i,i}^{(2)}}{h} - \epsilon_{11}^S\nabla^2\Phi^{(1)} + \frac{3\epsilon_{33}^S\Phi^{(1)}}{h^2} = \frac{3\sigma}{h} \quad (5.102)$$

where $i = 1$ and 2. Here, σ is the free charge per unit area on the electrode at $x_3 = +h$.

The factors K_1 to K_4 were determined by matching exact and approximate solutions for, among other things, the resonant frequency of thickness dilational resonance for an infinite electroded plate. The factor K_5 is introduced, as we have said, primarily to bring exact and approximate solutions for electric and piezoelectric effects into closer agreement.

It is well known that the separation of resonant and antiresonant frequencies in a piezoelectric resonator is an indicator of the strength of piezoelectric effects in that resonator.[131] As we have already used K_1 to K_4 to match the exact and approximate thickness-dilational resonant frequency of an infinite electroded plate, it thus seems logical to use K_5 for matching the exact and approximate antiresonant frequency of such a plate. Then, at least in the limit of an infinitely large electrode, the exact and approximate solutions will yield identical values for the piezoelectric strength of thickness-dilational resonance.

At the antiresonant frequency of an infinite electroded plate σ, but not $\Phi^{(1)}$, in Equations 5.79 and 5.102 is zero. All derivatives with respect to x_1 or x_2 are zero, however. We thus obtain from the second of Equations 5.79 and from 5.102 the following approximate expression for the thickness dilational antiresonant frequency

$$\begin{aligned}\Omega_{A(2)}^2 &= \frac{12}{\pi^2}\frac{K_1^2}{K_3^2}\frac{c_{33}^E}{c_{55}^E}\left(1 + \frac{K_5^2e_{33}^2}{c_{33}^E\epsilon_{33}^S}\right) \\ &= \frac{4\kappa^2}{\pi^2}\frac{c_{33}^D}{c_{55}^E}\left(1 + K_5^2\frac{k_t^2}{1 - k_t^2}\right) \end{aligned} \quad (5.103)$$

In obtaining this result, use has been made of Equation 5.94 and the definition of k_t^2, the thickness dilational coupling factor,

$$\frac{k_t^2}{1 - k_t^2} = \frac{e_{33}^2}{c_{33}^E\epsilon_{33}^S} \quad (5.104)$$

The approximate value of $\Omega^2_{A(2)}$ given by Equation 5.103 must be matched with the exact value,

$$\Omega^2_{A(2)} = c^D_{33}/c^E_{55} \tag{5.105}$$

This then yields the formula for K_5

$$K_5^2 = \frac{1 - k_t^2}{k_t^2}\left(\frac{\pi^2 - 4\kappa^2}{4\kappa^2}\right) \tag{5.106}$$

For PZT-5A, based on the coefficients appearing at the bottom of Table 5.1, K_5 is 0.901.

5.3.4. The Wave Equations in the Surrounding Unelectroded Region

The ceramic plate in the surrounding unelectroded region of Figure 5.19 has been specified to be electrically depolarized, and hence piezo-inactive. Consequently, the wave equations relevant to this region may be obtained from 5.79, the equations for the electroded region, by neglecting electric and piezoelectric effects,

$$\begin{aligned}
&\tfrac{1}{2}(c'_{11} + c'_{12})U^{(0)}_{i,ij} + c'_{66}\nabla^2 U^{(0)}_j + \frac{K'_1 c'_{13} U^{(1)}_{3,j}}{h} + \rho\omega^2 U^{(0)}_j = 0\\
&K'^2_2 c'_{55}\nabla^2 U^{(1)}_3 - \frac{3c'_{13}K'_1 U^{(0)}_{i,i}}{h} - \frac{3K'^2_1 c'_{33} U^{(1)}_3}{h^2}\\
&\qquad + \frac{3c'_{55}K'^2_2 U^{(2)}_{i,i}}{h} + \rho K'^2_3\omega^2 U^{(1)}_3 = 0\\
&c'_{66}\nabla^2 U^{(2)}_j + \frac{E''(1+\nu')}{2} U^{(2)}_{i,ij}\\
&\qquad - \frac{5}{h}[K'^2_2 c'_{55}(U^{(1)}_{3,j} + 3h^{-1}U^{(2)}_j)] + \rho K'^2_4\omega^2 U^{(2)}_j = 0
\end{aligned} \tag{5.107}$$

Here, we have defined

$$E'' = c'_{11} - c'^2_{13}/c'_{33} \tag{5.108}$$

and

$$\nu' = -s'_{12}/s'_{11} \tag{5.109}$$

In Equations 5.107 to 5.109, the c, K, s, ν, and E' are all primed to distinguish them from the previously used values relevant to the electroded area (see Equations 5.79 to 5.82).

As the unelectroded area is not piezoactive, it is unnecessary to use a correction factor K_5'. The other unelectroded area correction factors K_1' to K_4' are determined by applying Equations 5.92 to 5.99 in the piezoinactive limit,

$$\frac{K_1'}{K_3'} = \frac{\pi}{\sqrt{12}} \qquad \frac{K_2'}{K_4'} = \frac{\pi}{\sqrt{15}}$$

$$N_2' = \frac{3}{K_3'^2 \Omega'^3_{co(2)}} \left[\frac{4K_1'^2 c_{13}'^2}{\pi^2 c_{55}'} - \frac{\Omega'^4_{co(2)} K_2'^2}{3(4 - \Omega'^2_{co(2)})} \right]$$

$$= \left. \frac{d^2\Omega'}{dz'^2_{(2)}} \right|_{\Omega' = \Omega'_{co(2)}} \tag{5.110}$$

$$N_3' = \frac{1}{2} \left[\frac{E''}{K_4'^2 c_{55}'} - \frac{K_2'^2}{3K_1'^2 c_{33}' / (\pi^2 c_{55}') - K_3'^2} \right]$$

$$= \left. \frac{d^2\Omega'}{dz'^2_{(3)}} \right|_{\Omega' = \Omega'_{co(3)}}$$

where

$$\Omega'^2 = \left(\frac{2h\omega}{\pi} \right)^2 \frac{\rho}{c_{55}'}$$

$$\Omega'^2_{co(2)} = \frac{c_{33}'}{c_{55}'} = \left(\frac{2h\omega'_{co(2)}}{\pi} \right)^2 \frac{\rho}{c_{55}'} \tag{5.111}$$

$$\Omega'^2_{co(3)} = 4 = \left(\frac{2h\omega'_{co(3)}}{\pi} \right)^2 \frac{\rho}{c_{55}'}$$

For electrically depolarized PZT-5A, we have $c_{11}' = 13.56 \times 10^{10}$, $c_{12}' = 8.84 \times 10^{10}$, $c_{13}' = 8.59 \times 10^{10}$, and $c_{55}' = 2.91 \times 10^{10}$ in MKS units. This set of coefficients leads to values $N_2' = 8.626$, $N_3' = -6.858$, $K_1' = 0.982$, $K_2' = 1.510$, $K_3' = 1.083$, and $K_4' = 1.861$.

Note that $\Omega'_{co(2)}$ and $\Omega'_{co(3)}$ in Equations 5.111 represent the actual unelectroded region cutoff frequencies $\omega'_{co(m)}$ normalized into dimensionless form by c_{55}' rather than by c_{55}^E. This is in contrast to the normalized electroded region cutoff frequencies $\Omega_{co(2)}$ and $\Omega_{co(3)}$, appearing in Equations 5.90, 5.92, 5.95, 5.98, and 5.99.

Energy trapping may occur in the band between

$$\Omega = \Omega_{co(2)} \qquad (= 2.2658 \text{ for PZT-5A}) \tag{5.112}$$

and

$$\Omega = \Omega'_{co(2)} (c_{55}'/c_{55}^E)^{1/2} = (c_{33}'/c_{55}^E)^{1/2} \qquad (= 2.3452 \text{ for PZT-5A}) \tag{5.113}$$

In Equation 5.113, the factor $(c'_{55}/c^E_{55})^{1/2}$ is required to compensate for the earlier-mentioned use of c'_{55} rather than c^E_{55} in normalizing $\Omega'_{co(2)}$.

5.3.5. Numerical Curves for the Dispersion Relation

The first three branches of the dispersion curve $z^2_{(m)}(\Omega)$ where $m = 1, 2, 3$, for PZT-5A have been evaluated numerically, both by means of

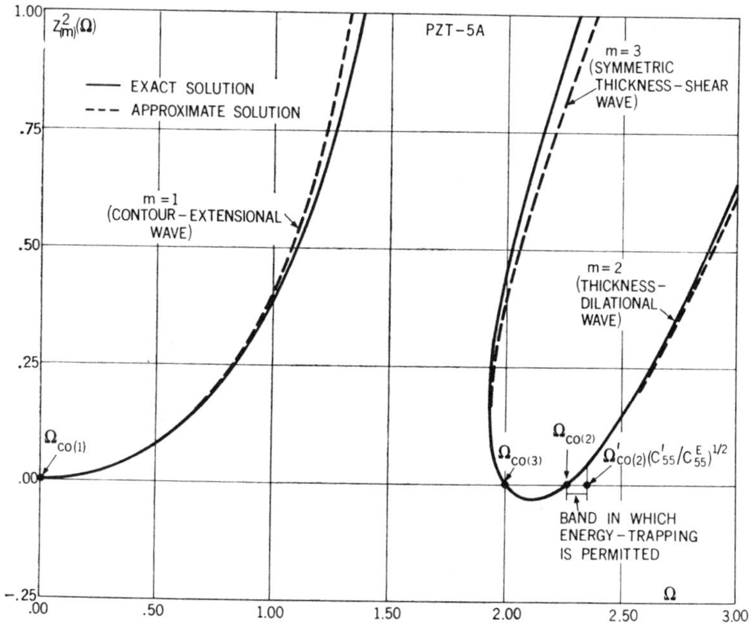

Figure 5.20. Exact (Reference 128) and approximate (Equations 5.84 and 5.88) dispersion curves for the first three straight-crested waves of dilation symmetry in an infinite, fully electroded ferroelectric ceramic plate. Curves are based on PZT-5A. Electrodes are assumed to be shorted together and grounded.

the exact procedure described in Reference 128 and approximately from Equation 5.88 with the K_1 to K_4 on the left side of Table 5.1. Results are compared graphically in Figure 5.20. It may be seen from this figure that agreement between the exact and approximate curves is generally good. More important, however, is the locally excellent agreement of the two curves for the branch of interest (the thickness-dilation branch) in the frequency band of interest (the band between the two cutoff frequencies). It is in this band and on this branch that energy trapping will occur.

5.3.6. Solution of the Wave Equations

It is considerably easier to treat the homogeneous equations associated with 5.79 than to treat 5.79 itself. These homogeneous equations approximately describe the electroded portion of sample shown in Figure 5.19 when the electrodes are shorted together and grounded. Consequently, the resulting solutions approximate the short-circuit eigenmodes of the sample. By means of Equation 4.40, we may afterwards use the solutions for these eigenmodes to determine the sample's electrical admittance at the electrodes.

Let us begin by writing Equations 5.79 in vector form with driving terms omitted,

$$\begin{aligned}
&c_{11}^E \operatorname{grad} \operatorname{div} \mathbf{U}^{(0)} - c_{66}^E \operatorname{curl} \operatorname{curl} \mathbf{U}^{(0)} \\
&\qquad + \frac{K_1 c_{13}^E}{h} \operatorname{grad} U_3^{(1)} + \rho\omega^2 \mathbf{U}^{(0)} = 0 \\
&K_2^2 c_{55}^E \nabla^2 U_3^{(1)} - \frac{3c_{13}^E K_1}{h} \operatorname{div} \mathbf{U}^{(0)} - \frac{3K_1^2 c_{33}^E}{h^2} U_3^{(1)} \\
&\qquad + \frac{3c_{55}^E K_2^2}{h} \operatorname{div} \mathbf{U}^{(2)} + \rho K_3^2 \omega^2 U_3^{(1)} = 0 \\
&E' \operatorname{grad} \operatorname{div} \mathbf{U}^{(2)} - c_{66}^E \operatorname{curl} \operatorname{curl} \mathbf{U}^{(2)} \\
&\qquad - \frac{5K_2^2}{h} c_{55}^E(\operatorname{grad} U_3^{(1)} + 3h^{-1}\mathbf{U}^{(2)}) + \rho K_4^2 \omega^2 \mathbf{U}^{(2)} = 0
\end{aligned} \tag{5.114}$$

All vectors appearing here are understood to have x_1 and x_2 components only.

These equations may be shown to possess solutions of the form[126]

$$\begin{aligned}
\mathbf{U}^{(0)} &= \sum_{m=1}^{3} A_m \operatorname{grad} \psi_m - A_4 \operatorname{curl} (\mathbf{i}_3 H_0) \\
U_3^{(1)} &= \sum_{m=1}^{3} A_m \alpha_m \psi_m \\
\mathbf{U}^{(2)} &= \sum_{m=1}^{3} A_m \beta_m \operatorname{grad} \psi_m - A_5 \operatorname{curl} (\mathbf{i}_3 H_2)
\end{aligned} \tag{5.115}$$

In these expressions, A_1 to A_5 are undetermined coefficients, while

$$\begin{aligned}
\alpha_m &= \frac{a_{11}}{a_{12}} \frac{z\pi}{2h} \bigg|_{z=z_{(m)}(\Omega)} \\
\beta_m &= \frac{a_{11}a_{23}}{a_{12}a_{33}} \bigg|_{z=z_{(m)}(\Omega)}
\end{aligned} \tag{5.116}$$

where the a_{ij} are from Equations 5.85 and 5.88. The potential functions ψ_1, ψ_2, ψ_3, H_0, and H_2 are required to obey

$$\nabla^2\psi_m + \frac{\pi^2 z^2_{(m)}(\Omega)}{4h^2}\psi_m = 0 \qquad m = 1, 2, 3$$

$$\nabla^2 H_0 + \frac{\Omega^2\pi^2 c^E_{55}}{4h^2 c^E_{66}} H_0 = 0 \tag{5.117}$$

$$\nabla^2 H_2 + \left(\frac{\Omega^2\pi^2 c^E_{55} K_4^2}{4h^2 c^E_{66}} - \frac{15K_2^2 c^E_{55}}{h^2 c^E_{66}}\right) H_2 = 0$$

Equations 5.115 represent the solutions to the approximate wave equations in the electroded region, Equations 5.79. It is also necessary to have the solutions for the unelectroded region approximate wave equations, Equations 5.107. These solutions are obtained by omitting piezoelectric effects in Equations 5.115 to 5.117:

$$\mathbf{U}^{(0)} = \sum_{m=1}^{3} A'_m \operatorname{grad} \psi'_m - A'_4 \operatorname{curl} (\mathbf{i}_3 H'_0)$$

$$U_3^{(1)} = \sum_{m=1}^{3} A'_m \alpha'_m \psi'_m \tag{5.118}$$

$$\mathbf{U}^{(2)} = \sum_{m=1}^{3} A'_m \beta'_m \operatorname{grad} \psi'_m - A'_5 \operatorname{curl} (\mathbf{i}_3 H'_2)$$

where A'_1 to A'_5 are five more undetermined coefficients. In addition, α'_m and β'_m are

$$\alpha'_m = \left.\frac{a'_{11}}{a'_{12}}\frac{z\pi}{2h}\right|_{z=z'_{(m)}(\Omega')}$$

$$\beta'_m = \left.\frac{a'_{11}a'_{23}}{a'_{12}a'_{33}}\right|_{z=z'_{(m)}(\Omega')} \tag{5.119}$$

where a'_{ij} are found from Equations 5.85 and 5.88 again by omitting piezoelectric effects,

$$a'_{11} = z^2 c'_{11}/c'_{55} - \Omega'^2$$

$$a'_{12} = 2K'_1 c'_{13} z/(\pi c'_{55})$$

$$a'_{22} = z^2 K_2'^2/3 + 4K_1'^2 c'_{33}/(\pi^2 c'_{55}) - \Omega'^2 K_3'^2/3 \tag{5.120}$$

$$a'_{23} = -2K_2'^2 z/\pi$$

$$a'_{33} = 12K_2'^2/\pi^2 + E'' z^2/(5c'_{55}) - \Omega'^2 K_4'^2/5$$

The quantity Ω' appearing here has been defined in Equation 5.111 by reducing Equation 5.87 to piezoinactive form,

$$\Omega' = \frac{2h\omega}{\pi}\sqrt{\frac{\rho}{c'_{55}}} \tag{5.121}$$

so that

$$\Omega' = (c_{55}^E/c'_{55})^{1/2}\Omega \tag{5.122}$$

In addition, $z'_{(m)}(\Omega')$ are the roots of

$$\begin{vmatrix} a'_{11} & a'_{12} & 0 \\ a'_{12} & a'_{22} & a'_{23} \\ 0 & a'_{23} & a'_{33} \end{vmatrix} = 0 \tag{5.123}$$

The potential functions ψ'_1, ψ'_2, ψ'_3, H'_0, H'_2 appearing in Equations 5.118 must satisfy

$$\nabla^2\psi'_m + \frac{\pi^2 z'^2_{(m)}(\Omega')}{4h^2}\psi'_m = 0 \qquad m = 1, 2, 3$$

$$\nabla^2 H'_0 + \frac{\Omega'^2\pi^2 c'_{55}}{4h^2 c'_{66}} H'_0 = 0 \tag{5.124}$$

$$\nabla^2 H'_2 + \left(\frac{\Omega'^2\pi^2 c'_{55}K'^2_4}{4h^2 c'_{66}} - \frac{15K'^2_2 c'_{55}}{h^2 c'_{66}}\right) H'_2 = 0$$

5.3.7. *Boundary Conditions at the Electrode Edge*

In the solutions, Equations 5.115 and 5.118, there are ten undetermined coefficients, A_1 to A_5 and A'_1 to A'_5. These ten coefficients are evaluated by applying certain continuity conditions at the electrode boundary.

Let us express the unit normal vector of the boundary as **n** and the unit tangential vector as **ν**. (Both **n** and **ν** are understood to lie in the plane of the plate.) Then we should require the five displacements, $U_n^{(0)}$, $U_\nu^{(0)}$, $K_3U_3^{(1)}$, $K_4U_n^{(2)}$, and $K_4U_\nu^{(2)}$ to be continuous at the electrode boundary. Additionally, if we expand stress in the plate as[126]

$$T_{ij} = \sum_{\alpha=0}^{\infty} \tfrac{1}{2}(2\alpha + 1)T_{ij}^{(\alpha)}(x_1,x_2)P_\alpha(x_3/h) \tag{5.125}$$

we should like $T_{nn}^{(0)}$, $T_{nn}^{(2)}$, $T_{n\nu}^{(0)}$, $T_{n\nu}^{(2)}$, and $T_{n3}^{(1)}$ to be continuous. Subject to definition of Equation 5.125, one may express these $T_{ij}^{(\alpha)}$ as[126]

$$
\begin{aligned}
T_{nn}^{(0)} &= 2(c_{11}^E U_{n,n}^{(0)} + c_{12}^E U_{\nu,\nu}^{(0)} + K_1 C_{13}^E h^{-1} U_3^{(1)}) \\
\text{or} \quad & 2(c_{11}' U_{n,n}^{(0)} + c_{12}' U_{\nu,\nu}^{(0)} + K_1' c_{13}' h^{-1} U_3^{(1)}) \\
T_{nn}^{(2)} &= 2E'(U_{n,n}^{(2)} + \nu^D U_{\nu,\nu}^{(2)})/5 \\
\text{or} \quad & 2E''(U_{n,n}^{(2)} + \nu' U_{\nu,\nu}^{(2)})/5 \\
T_{n\nu}^{(0)} &= 2c_{66}^E(U_{n,\nu}^{(0)} + U_{\nu,n}^{(0)}) \\
\text{or} \quad & 2c_{66}'(U_{n,\nu}^{(0)} + U_{\nu,n}^{(0)}) \\
T_{n\nu}^{(2)} &= 2c_{66}^E(U_{n,\nu}^{(2)} + U_{\nu,n}^{(2)})/5 \\
\text{or} \quad & 2c_{66}'(U_{n,\nu}^{(2)} + U_{\nu,n}^{(2)})/5 \\
T_{n3}^{(1)} &= 2K_2^2 c_{55}^E(U_{3,n}^{(1)} + 3h^{-1}U_n^{(2)})/3 \\
\text{or} \quad & 2K_2'^2 c_{55}'(U_{3,n}^{(1)} + 3h^{-1}U_n^{(2)})/3
\end{aligned}
\qquad (5.126)
$$

Continuity of these five stresses and five displacements provides us with the necessary ten conditions on the ten A_m and A_m'.

5.3.8. *Application to Circular Electrodes*

Let us now illustrate how the concepts presented in this section are applied to the specific example in which the electroded region in Figure 5.19 is a circle of radius a. We shall assume the electrodes are centered at the origin of the coordinate system, so that

$$\frac{\partial}{\partial \theta} \equiv 0 \qquad (5.127)$$

Under these conditions, the electroded-region potential functions appearing in the differential equations of 5.117 have standing-wave-type Bessel function solutions,

$$
\begin{aligned}
\psi_m &= J_0\left(\frac{\pi z_{(m)}(\Omega) r}{2h}\right) \\
H_0 &= J_0\left(\frac{\pi \Omega r}{2h}\sqrt{\frac{c_{55}^E}{c_{66}^E}}\right) \\
H_2 &= J_0\left(\frac{\gamma r}{h}\right)
\end{aligned}
\qquad (5.128)
$$

where

$$\gamma^2 = \frac{c_{55}^E}{c_{66}^E}\left(\frac{\Omega^2\pi^2K_4^2}{4} - 15K_2^2\right) \tag{5.129}$$

Outside the electroded region, the potential function differential equations also have Bessel function solutions. However, in this case the relevant Bessel functions must be either outgoing traveling waves or evanescent waves,

$$\psi_m' = H_0^{(1)}\left(\frac{\pi z_{(m)}'(\Omega')r}{2h}\right) \to \frac{2}{\pi}\sqrt{\frac{h}{z_{(m)}'(\Omega')r}} \exp\left[j\left(\frac{\pi z_m'(\Omega')r}{2h} - \frac{\pi}{4}\right)\right]$$

$$H_0' = H_0^{(1)}\left(\frac{\pi\Omega' r}{2h}\sqrt{\frac{c_{55}'}{c_{66}'}}\right) \to \frac{2}{\pi}\sqrt{\frac{h}{\Omega' r}}\sqrt[4]{\frac{c_{66}'}{c_{55}'}} \exp\left[j\left(\frac{\pi\Omega' r}{2h}\sqrt{\frac{c_{55}'}{c_{66}'}} - \frac{\pi}{4}\right)\right]$$

$$H_2' = H_0^{(1)}\left(\frac{\gamma' r}{h}\right) \to \sqrt{\frac{2h}{\pi\gamma' r}} \exp\left[j\left(\frac{\gamma' r}{n} - \frac{\pi}{4}\right)\right] \tag{5.130}$$

where

$$\gamma'^2 = \frac{c_{55}'}{c_{66}'}\left(\frac{\Omega'^2\pi^2K_4'^2}{4} - 15K_2'^2\right) \tag{5.131}$$

Energy trapping is permitted in the frequency band specified by Equations 5.112 and 5.113,

$$2.2658 < \Omega < 2.3452 \tag{5.132}$$

for PZT-5A. In this band, $z_{(1)}(\Omega)$, $z_{(1)}'(\Omega)$, and $z_{(2)}(\Omega)$ will always be real. In addition, $z_{(2)}'(\Omega)$ will be imaginary. For many materials, including PZT-5A, $z_{(3)}(\Omega)$, $z_{(3)}'(\Omega)$, γ, and γ' will also be real (see Figure 5.20). There is a possibility that some or all of these last four quantities can be imaginary, however. This phenomenon may occur if c_{33}'/c_{55}^E or c_{33}^E/c_{55}^E is near or less than 4, so that $\Omega_{co(3)}'(c_{55}'/c_{55}^E) > \Omega_{co(2)}$ or $\Omega_{co(3)} > \Omega_{co(2)}$ (see 5.92, 5.95, 5.111, and Figure 5.20). For our present example, we shall assume this does not happen. That means all the wavenumbers except $z_{(2)}'(\Omega)$ are real, and $z_{(2)}'(\Omega)$ is imaginary.

Then the ψ_1', ψ_3', H_0', and H_2' solutions in Equations 5.130 represent outward traveling cylindrical waves which can carry net energy away from the electrodes. In other words, in this case, energy may not be perfectly trapped in the thickness-dilation wave at the electrodes. If any mode conversion into the ψ_1', ψ_3', H_0', or H_2' waves occurs at the electrode edge, that energy leaks off. This situation is analogous to that described in the previous section for the case of thickness-twist waves with $n > 1$. However, thickness-dilation trapped energy configurations differ from thickness-twist configurations in that at least the fundamental thickness-twist wave may be perfectly trapped. The present section, which applies only to the fundamental thickness-dilational wave, indicates that even the fundamental wave of this type is not necessarily perfectly trapped.

The $z'_{(2)}(\Omega')$ wavenumber, as we have said, will always be imaginary if energy trapping is possible,

$$z'_{(2)}(\Omega') = iy'_{(2)}(\Omega') \tag{5.133}$$

Consequently ψ'_2 is always an evanescent Bessel function,

$$\begin{aligned}\psi'_2(\Omega') &= H_0^{(1)}\left(\frac{i\pi y'_{(2)}(\Omega')r}{2h}\right)\\ &= -\frac{2i}{\pi}K_0\left(\frac{\pi y'_{(2)}(\Omega')r}{2h}\right)\\ &\to -\frac{2i}{\pi}\sqrt{\frac{h}{y'_{(2)}(\Omega')r}}\exp\left[-\pi y'_{(2)}(\Omega')r/(2h)\right]\end{aligned} \tag{5.134*}$$

It is because $z_{(2)}(\Omega)$ is real while $z'_{(2)}(\Omega')$ is imaginary that the evanescent condition, and hence thickness-dilation energy trapping, is permissible for $2.2658 < \Omega < 2.3452$.

If the potential solutions of Equations 5.128 and 5.130 are actually substituted into Equations 5.115, 5.118, and 5.126, we get the following dimensionless† conditions on the A_m and A'_m:

Continuity of $U_r^{(0)}$

$$\sum_{m=1}^{3} A_m\left[z_{(m)}(\Omega)J_1\left(\frac{\pi z_{(m)}(\Omega)B}{2}\right)\right] - \sum_{m=1}^{3} A'_m\left[z'_{(m)}(\Omega')H_1^{(1)}\left(\frac{\pi z'_{(m)}(\Omega')B}{2}\right)\right] = 0 \tag{5.135}$$

* Our reason for selecting $e^{-j\omega t}$ rather than $e^{+j\omega t}$ time dependence in this chapter becomes significant at this point. Had we made the alternative selection, the outward-traveling radial waves would be $H_0^{(2)}(\pi z'_{(2)}(\Omega')r/2h)$. These $H_0^{(2)}$ functions, unlike the $H_0^{(1)}$, are not simply related to the K_0 or evanescent functions.

† If $(\alpha_m h)$, $(\alpha'_m h)$, and ratios of c_{ij}'s are considered to be single quantities, everything appearing in the following continuity equations is dimensionless.

Continuity of $K_3 U_3^{(1)}$

$$\sum_{m=1}^{3} A_m\left[(\alpha_m h)K_3 J_0\left(\frac{\pi z_{(m)}(\Omega)B}{2}\right)\right] - \sum_{m=1}^{3} A'_m\left[(\alpha'_m h)K'_3 H_0^{(1)}\left(\frac{\pi z'_{(m)}(\Omega')B}{2}\right)\right] = 0 \quad (5.136)$$

Continuity of $K_4 U_r^{(2)}$

$$\sum_{m=1}^{3} A_m\left[K_4 \beta_m z_{(m)}(\Omega) J_1\left(\frac{\pi z_{(m)}(\Omega)B}{2}\right)\right] - \sum_{m=1}^{3} A'_m\left[K'_4 \beta'_m z'_{(m)}(\Omega') H_1^{(1)}\left(\frac{\pi z'_{(m)}(\Omega')B}{2}\right)\right] = 0 \quad (5.137)$$

Continuity of $T_{rr}^{(0)}$

$$\sum_{m=1}^{3} A_m\left[\frac{c_{11}^E}{c_{55}^E}\left(\frac{\pi z_{(m)}(\Omega)}{2}\right)^2 J'_1\left(\frac{\pi z_{(m)}(\Omega)B}{2}\right) + \frac{c_{12}^E}{c_{55}^E}\cdot\frac{1}{B}\left(\frac{\pi z_{(m)}(\Omega)}{2}\right) J_1\left(\frac{\pi z_{(m)}(\Omega)B}{2}\right) - \frac{(\alpha_m h)K_1 c_{13}^E}{c_{55}^E} J_0\left(\frac{\pi z_{(m)}(\Omega)B}{2}\right)\right]$$
$$- \sum_{m=1}^{3} A'_m\left[\frac{c'_{11}}{c_{55}^E}\left(\frac{\pi z'_{(m)}(\Omega')}{2}\right)^2 H_1^{(1)\prime}\left(\frac{\pi z'_{(m)}(\Omega')B}{2}\right) + \frac{c'_{12}}{c_{55}^E}\cdot\frac{1}{B}\left(\frac{\pi z'_{(m)}(\Omega')}{2}\right) H_1^{(1)}\left(\frac{\pi z'_{(m)}(\Omega')B}{2}\right) - \frac{(\alpha'_m h)K'_1 c'_{13}}{c_{55}^E} H_0^{(1)}\left(\frac{\pi z'_{(m)}(\Omega')B}{2}\right)\right] = 0 \quad (5.138)$$

Continuity of $T_{rr}^{(2)}$

$$\sum_{m=1}^{3} A_m\left[\frac{E'}{c_{33}^E}\beta_m z_{(m)}(\Omega)\left\{\frac{\nu^D}{B} J_1\left(\frac{\pi z_{(m)}(\Omega)B}{2}\right) + \frac{\pi z_{(m)}(\Omega)}{2} J'_1\left(\frac{\pi z_{(m)}(\Omega)B}{2}\right)\right\}\right]$$
$$- \sum_{m=1}^{3} A'_m\left[\frac{E''}{c_{55}^E}\beta'_m z'_{(m)}(\Omega')\left\{\frac{\nu'}{B} H_1^{(1)}\left(\frac{\pi z'_{(m)}(\Omega')B}{2}\right) + \frac{\pi z'_{(m)}(\Omega')}{2} H_1^{(1)\prime}\left(\frac{\pi z'_{(m)}(\Omega')B}{2}\right)\right\}\right] = 0 \quad (5.139)$$

Continuity of $T_{r3}^{(1)}$

$$\sum_{m=1}^{3} A_m\left[K_2^2 z_{(m)}(\Omega)(\alpha_m h + 3\beta_m)J_1\left(\frac{\pi z_{(m)}(\Omega)B}{2}\right)\right]$$
$$-\sum_{m=1}^{3} A'_m\left[K_2'^2 \frac{c'_{55}}{c_{55}^E} z'_{(m)}(\Omega')(\alpha'_m h + 3\beta'_m)H_1^{(1)}\left(\frac{\pi z'_{(m)}(\Omega')B}{2}\right)\right] = 0$$

(5.140)

In these continuity equations, B represents the electrode diameter to plate thickness ratio,

$$B = 2a/(2h) \qquad (5.141)$$

One may observe that Equations 5.135 to 5.140 constitute only six equations in six unknowns, whereas we commented in Section 5.3.7 that actually ten unknowns and ten boundary conditions should be considered. However, in this particular example the ten equations in ten unknowns decouple into

1. the six equations in A_1 to A_3 and A'_1 to A'_3 given by Equation 5.135 to 5.140;
2. two equations in A_4 and A'_4 representing continuity of $U_\theta^{(0)}$ and $T_{r\theta}^{(0)}$;
3. two equations in A_5 and A'_5 representing continuity of $K_4U_\theta^{(2)}$ and $T_{r\theta}^{(2)}$.

The second two sets of decoupled equations characterize uniform shear and symmetric thickness-twist motions, which are not of interest to us as they are piezoinactive. Consequently, in this example at least, the ten equations in ten unknowns become reduced to six equations in six unknowns.

Physically, this lack of coupling means that mode conversion from the thickness-dilation wave into the H'_0 and H'_2 waves does not occur at the electrode edge. Consequently, energy leakage cannot occur *via* these waves, even though they would propagate outside the electroded region if somehow they were excited. Mode conversion into the propagating ψ'_1 and ψ'_3 waves will occur, however, so that we still do not obtain ideal trapping.

Generally speaking, mode conversion from dilational to uniform shear (H'_0) and symmetric thickness-twist (H'_2) waves will not occur at an interface whenever the wavenumber vectors are normal to the interface: such is the case here. This is the general condition for reducing the number of coupled equations from ten to six.

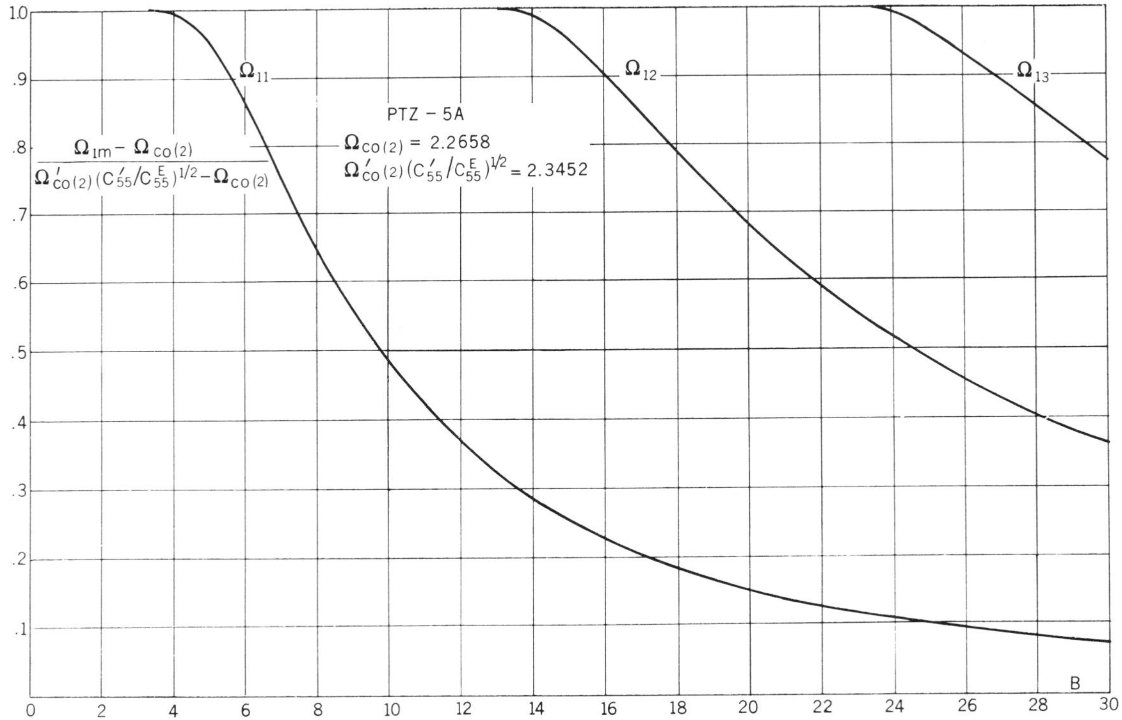

Figure 5.21. Trapped-energy resonance spectrum of thickness-dilational waves for a PZT-5A plate. This figure is based on a circular electrode trapping region with B = electrode radius/plate thickness. Energy leakage or imaginary components of Ω_{1m} are omitted.

Equations 5.135 to 5.140 will have nontrivial solutions only for those discrete values of Ω at which the determinant of the associated matrix vanishes. These eigenvalues, which we shall designate Ω_{1n} to correspond with the notation of the previous section, represent the normalized resonant frequencies of energy trapping.

The spectrum for Ω versus B based on PZT-5A as computed from Equations 5.135 to 5.140 is illustrated in Figure 5.21. In this figure, the small imaginary components of Ω_{1n} associated with energy leakage have been omitted; as in the thickness-twist example, loss because of energy leakage is negligible compared to the loss associated with material dissipation. (This is true at least for PZT-5A.) Data is actually presented in Figure 5.21 in the form $[\Omega_{1n} - \Omega_{co(2)}]/[\Omega'_{co(2)}(c'_{55}/c^E_{55})^{1/2} - \Omega_{co(2)}]$. This expression is the fractional spacing of Ω_{1n} between the upper and lower limits of the band in which energy trapping is permitted.

Only one really significant difference occurs between this thickness-dilation spectrum and the thickness-twist spectra shown in Figures 5.10 and 5.13: There are no trapped resonances at all in the present case when the electrode is too small ($B < 2.8$). The thickness-twist spectra show at least one resonance for all values of B. This difference is apparently attributable to the geometrical dissimilarities in the two examples: In the one-dimensional thickness-twist case, the electroded area is actually infinite for any value of B. Here the circular electrodes, obviously, are finite for all B.

At the resonant frequencies Ω_{1n}, the A_m and A'_m are determined by Equations 5.135 to 5.140 only to within a multiplicative constant. That constant itself is found by applying the normalization condition, Equations 4.30 and 4.32:

$$\begin{aligned}1 &= \int_0^\infty dr \int_{-h}^{h} dx_3 \int_0^{2\pi} d\theta \; r\rho\big([K_3 P_1(x_3/h) U_3^{(1)}(r)]^2 \\ &\quad + [P_0(x_3/h)U_r^{(0)}(r) + K_4 P_2(x_3/h) U_r^{(2)}(r)]^2\big) \\ &= 2\pi(\tfrac{2}{3}I_1 + I_0 + \tfrac{2}{5}I_2) \qquad (5.142)\end{aligned}$$

where

$$\begin{aligned}I_1 &= \int_0^\infty r[K_3\sqrt{\rho h}\; U_3^{(1)}(r)]^2\, dr \\ &= \int_0^a r\left[\sum_{m=1}^{3} (\sqrt{\rho h}\; A_m)\alpha_m K_3 J_0\left(\frac{\pi z_{(m)}(\Omega_{1n})r}{2h}\right)\right]^2 dr \\ &\quad + \int_a^\infty r\left[\sum_{m=1}^{3} (\sqrt{\rho h}\; A'_m)\alpha'_m K'_3 H_0^{(1)}\left(\frac{\pi z'_{(m)}(\Omega_{1n})r}{2h}\right)\right]^2 dr\end{aligned}$$

$$I_0 = \int_0^\infty r[\sqrt{\rho h}\; U_r^{(0)}(r)]^2\, dr$$

$$= \int_0^a r\left[\sum_{m=1}^{3} (\sqrt{\rho h}\; A_m)\, \frac{\pi z_{(m)}(\Omega_{1n})}{2h}\, J_1\!\left(\frac{\pi z_{(m)}(\Omega_{1n}) r}{2h}\right)\right]^2 dr$$

$$+ \int_a^\infty r\left[\sum_{m=1}^{3} (\sqrt{\rho h}\; A'_m)\, \frac{\pi z'_{(m)}(\Omega_{1n})}{2h}\, H_1^{(1)}\!\left(\frac{\pi z'_{(m)}(\Omega_{1n}) r}{2h}\right)\right]^2 dr$$

$$I_2 = \int_0^\infty r[K_4\sqrt{\rho h}\; U_r^{(2)}(r)]^2\, dr$$

$$= \int_0^a r\left[\sum_{m=1}^{3} (\sqrt{\rho h}\; A_m)\, \frac{\pi \beta_m z_{(m)}(\Omega_{1n})}{2h}\, K_4 J_1\!\left(\frac{\pi z_{(m)}(\Omega_{1n}) r}{2h}\right)\right]^2 dr$$

$$+ \int_a^\infty r\left[\sum_{m=1}^{3} (\sqrt{\rho h}\; A'_m)\, \frac{\pi \beta'_m z'_{(m)}(\Omega_{1n})}{2h}\right.$$

$$\left. \times\; K'_4 H_1^{(1)}\!\left(\frac{\pi z'_{(m)}(\Omega_{1n}) r}{2h}\right)\right]^2 dr \qquad (5.143)$$

These integrals, which are quadratic in A_m and A'_m, may be reduced to widely used integrals appearing in many integral tables. The only special consideration to be taken in their evaluation is that the real traveling-wave wavenumbers $z'_{(1)}(\Omega_{1n})$ and $z'_{(3)}(\Omega_{1n})$ should be given a small positive imaginary component. Physically this modification represents the effect of material dissipation; mathematically it causes the three integrals over $a \leq r < \infty$ to converge at $r \to \infty$.

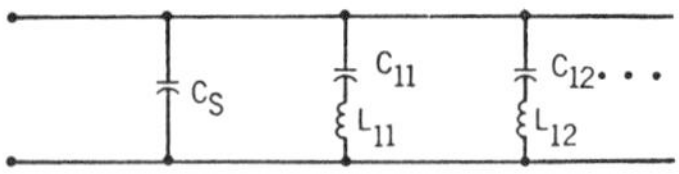

Figure 5.22. Equivalent circuit for the circular electrode dilation-wave trapped energy resonator. This circuit is valid at frequencies below the lowest thickness-dilation overtone resonance.

Once the dimensionless and normalized quantities $\sqrt{\rho h}\, A_m$ and $\sqrt{\rho h}\, A'_m$ have been evaluated by means of Equations 5.135 to 5.143, one can use the techniques of Section 4.2 to determine the admittance and equivalent circuit of the circular electrode pair. In particular, at frequencies below the thickness-dilation overtones, the circular electrodes are representable by the circuit shown in Figure 5.22.

This equivalent circuit corresponds to the admittance

$$Y = j\omega C_S + j\omega \sum_n \frac{(Q^{(1n)})^2}{\omega_{1n}^2 - \omega^2} \qquad (5.144)$$

which we obtain from Equation 4.40. Consequently, the lumped circuit elements appearing in Figure 5.22 are specified by

$$L_{1n}C_{1n} = \omega_{1n}^{-2} \tag{5.145}$$

and

$$C_{1n} = \left(\frac{Q^{(1n)}}{\omega_{1n}}\right)^2 \tag{5.146}$$

Additionally, C_S is the infinite-frequency or mechanically clamped capacitance of the electrode pair,

$$C_S = \epsilon_{33}^S \pi a^2/2h \tag{5.147}$$

and $Q^{(1n)}$ is the charge on the top electrode because of the normalized $(1n)$th trapped short-circuit resonance,

$$\begin{aligned} Q^{(1n)} &= -2\pi \int_0^a r D_3 \Big|_{\substack{\omega=\omega_{1n}\\ x_3=h}} dr \\ &= -2\pi \int_0^a r \Bigg[e_{31}\left(\frac{\partial U_r^{(0)}(r)}{\partial r} + \frac{U_r^{(0)}(r)}{r}\right) \\ &\quad + K_1K_5 e_{33} h^{-1} U_3^{(1)}(r)\Bigg]_{\substack{\omega=\omega_{1n}\\ x_3=h}} dr \end{aligned} \tag{5.148}$$

If $Q^{(1n)}$ is evaluated by substituting $U_r^{(0)}$ and $U_3^{(1)}$ from Equations 5.115 and 5.128, we find

$$\begin{aligned} Q^{(1n)} &= \frac{e_{33}B}{\sqrt{\rho h}} \sum_{n=1}^{3} (\sqrt{\rho h}\, A_m) J_1\left(\frac{\pi z_{(m)}(\Omega_{1n})B}{2}\right) \\ &\quad \cdot \left(\frac{\pi^2 e_{31} z_{(m)}(\Omega_{1n})}{e_{33}} - \frac{4(h\alpha_m)K_1K_5}{z_{(m)}(\Omega_{1n})}\right) \end{aligned} \tag{5.149}$$

In this equation, the $(\sqrt{\rho h}\, A_m)$ are understood to be normalized by Equation 5.142.

The piezoelectric strength of resonance $(1n)$ is proportional to the dynamic capacitance of that resonance C_{1n}. This capacitance may be obtained from Equations 5.146 and 5.149. A high C_{1n} indicates that the electrode admittance near ω_{1n} is high.

However, it is inconvenient to present numerical data on C_{1n} directly because it is not dimensionless. Numerical data on the strength of resonance $(1n)$ is better given as the normalized dynamic capacitance.

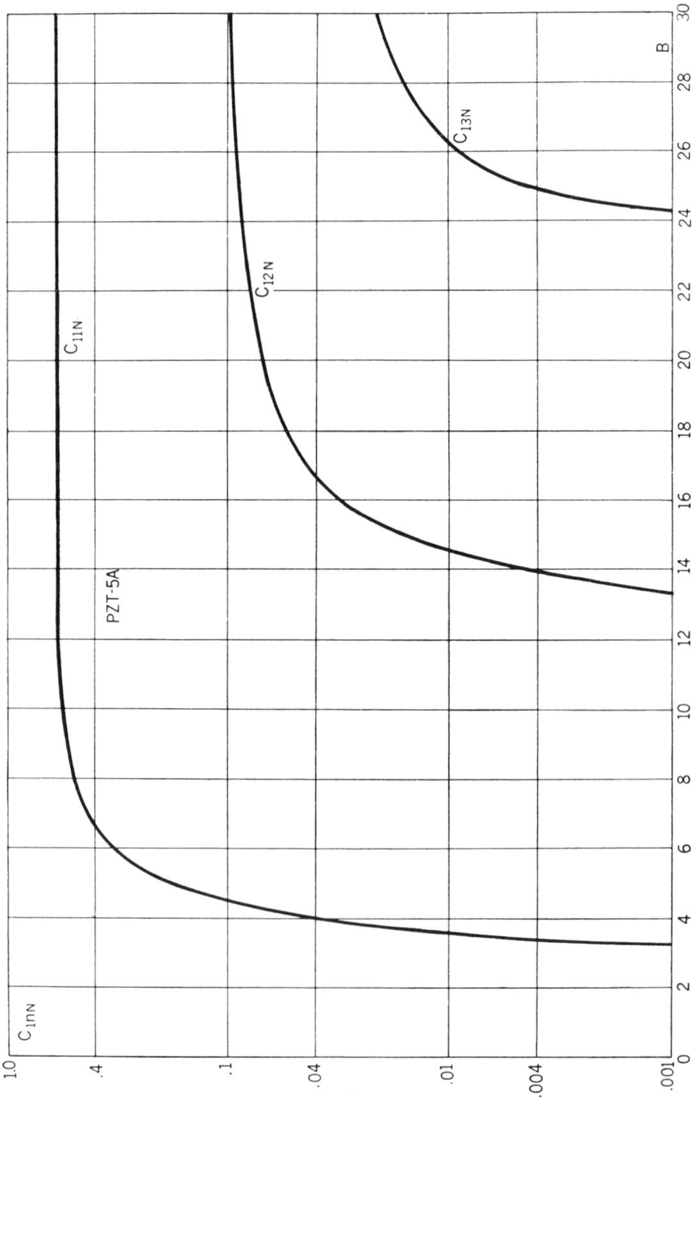

Figure 5.23. Normalized dynamic capacitances for the trapped thickness-dilational resonances of a PZT-5A plate with a circular electrode region.

We shall define this dimensionless quantity as

$$C_{1nN} = \frac{C_{1n}}{(\pi a^2/2h) \cdot \epsilon_{33}^S \cdot k_t^2/(1 - k_t^2)} \tag{5.150}$$

Once values for C_{1nN} are given, it is immediately possible to determine C_{1n} by inverting Equation 5.150, provided the plate dimensions and material coefficients are known.

If we combine Equations 5.146, 5.149, and 5.150 with Equation 5.87, we obtain the following expression for C_{1nN},

$$C_{1nN} = \frac{8}{\pi^3} \frac{c_{33}^E}{c_{55}^E} \frac{1}{\Omega_{1n}^2} \left\{ \sum_{m=1}^{3} (\sqrt{\rho h}\, A_m) \cdot J_1\left(\frac{\pi z_{(m)}(\Omega_{1n}) B}{2}\right) \left(\frac{\pi^2 e_{31} z_{(m)}(\Omega_{1n})}{e_{33}} - \frac{4(h\alpha_m) K_1 K_5}{z_{(m)}(\Omega_{1n})}\right) \right\}^2 \tag{5.151}$$

Values for C_{1nN} as computed from Equation 5.151 are shown in Figure 5.23. This figure is based on PZT-5A.

5.4. Effect of a High-Impedance Source on Energy-Trapping Resonators

5.4.1. General Remarks

The techniques described previously in this chapter have led to input admittance or equivalent circuit representations of energy-trapping piezoelectric resonators. These representations must, of course, be valid independently of the nature of the externally applied driving source. However, in some applications, the overall system response including both resonator and driving source may be very different from the response of the resonator alone. In these cases, the correlation between the system response and the resonator response will not be superficially recognizable. This is especially likely to be true when the source impedance is high compared to the resonator impedance at resonance.

The reason for these effects is that, in general, a finite source impedance results in system resonant frequencies somewhat above the short-circuit resonator resonances (see Figure 5.24). For noninductive sources, this resonant frequency increase is monotonic with source impedance: At zero source impedance, of course, the system resonances coincide with the short-circuit resonator resonances. However, as

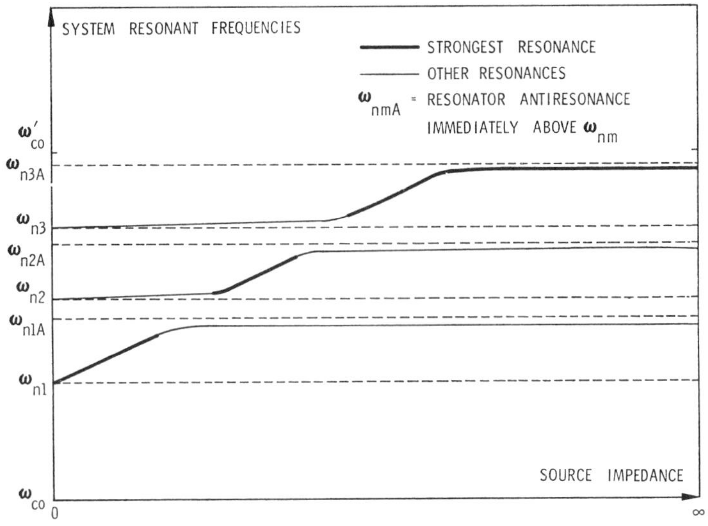

Figure 5.24. Effect on system resonant frequencies of a source impedance increasing from zero (voltage source) towards infinity (current source).

source impedance increases from zero toward infinity, the system resonant frequencies asymptotically increase toward the antiresonant frequencies of the resonator alone (see Figure 5.24). When several frequencies occur in close proximity, as in the case of a trapped energy device with large electrodes, the asymptotic approach can be somewhat complex. This complexity develops because a theorem analogous to that discussed in Section 4.4 applies: System resonant frequencies when plotted against source impedance are not allowed to cross (again see Figure 5.24).

For a nonrigorous illustration of what can happen, let us consider the thickness-twist energy-trapping configuration discussed in Section 5.2. There, both the trapping electroded region and the surrounding unelectroded region were fully poled ferroelectric ceramic (see Figure 5.3). Let us assume the electroded region is large enough to permit the existence of more than one trapped resonance in each anharmonic series. As Figures 5.16, 5.17, and 5.18 illustrate, the first of the system anharmonic resonances, $m = 1$, is the strongest piezoelectrically for each value of n when source impedance is low. (That is, the first resonance, $m = 1$, has the largest normalized dynamic capacitance C_{nm} for each value of n.) However, as source impedance increases, the system resonance with the strongest response does not remain $m = 1$

but rather changes monotonically to higher m values (again see Figure 5.24). For very large source impedances, the *last* not the first member of the anharmonic series is the most strongly excited. In this case, the spurious responses will occur at frequencies *below*, not above, the main response. Consequently, they may not be recognized as components of the anharmonic series.

A more detailed view of the effect of a very large source impedance on the configuration in Figure 5.3 may be presented as follows: The cutoff frequency of the first thickness-twist wave in the unelectroded region is ω'_{co}, the resonant frequency with $D = 0$ of a thickness acoustic standing half-wave in the unelectroded region. However, the cutoff frequency of that wave in the electroded region when the electrodes are connected to a very high impedance approaches ω''_{co}, the resonant frequency, also with $D = 0$, of a thickness acoustic standing half-wave in the electroded region. Note that ω''_{co} is not equal to ω_{co}. In fact if Rh, the electrode thickness, is negligible, we have

$$\omega''_{co} = \omega'_{co} \tag{5.152}$$

In other words, as source impedance approaches infinity, the system resonant frequency corresponding to the trapped mode which has no nodal lines approaches ω'_{co}, the upper frequency limit of the band in which energy trapping is permitted. This no-nodal-lines mode will, of course, be the strongest, as all other modes will have nodes and thus will not couple to the electrodes with uniform phase. As the system resonant frequency of the strongest mode progressively crowds closer and closer to ω'_{co} with increasing source impedance, all other members of the system anharmonic series are forced to occur at lower frequencies than that of the strongest mode. This forcing of the weaker resonances to lower frequencies must, of course, develop without any system resonant frequencies crossing each other. An illustration of this type of transition in which no resonance crossings occur appears in Figure 5.24.

If these remarks have served to confuse the reader, we may remind him of the reassurance stated at the beginning of this section: There is nothing actually wrong with using the mathematical techniques described in Sections 5.2 and 5.3 for energy-trapped devices, even if the device is to be driven by a high impedance source. The caution is that these techniques when applied to the high-source-impedance case will correctly predict some effects which are very unexpected or non-intuitive.

5.4.2. A Specific Example

The concepts just discussed in Section 5.4.1 will now be illustrated with a numerical design problem. Bandpass filters may be constructed from two cascaded circuits each employing one ferroelectric ceramic resonator.[133] One circuit is arranged so as to give maximum rejection just above the desired bandpass, and the other is arranged so as to give maximum rejection just below the bandpass. The circuit which rejects frequencies just below the bandpass usually requires a resonator driven by a high impedance source. A typical example of this type of circuit is illustrated in Figure 5.25.

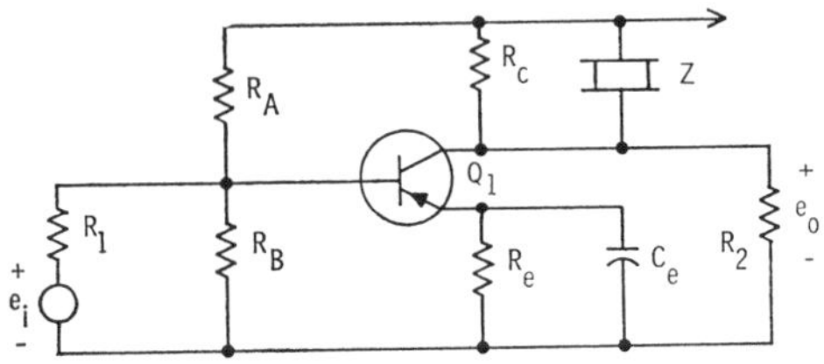

Figure 5.25. A circuit with maximum rejection at frequencies just below its bandpass.[133] Note that the piezoelectric resonator is located at a high impedance point.

If we represent the transistor in Figure 5.25 by its h parameters,

$$\begin{aligned} V_{be} &= I_b h_{ie} + V_{ce} h_{re} \doteq I_b h_{ie} \\ I_c &= I_b h_{fe} + V_{ce} h_{oe} \doteq I_b h_{fe} \end{aligned} \tag{5.153}$$

we can express the gain of the circuit in question as

$$G = \frac{e_o}{e_i} \doteq -\frac{Z_2 h_{fe}}{Z_1 + h_{ie}} \cdot \frac{Z_1}{R_1} \tag{5.154}$$

where

$$\begin{aligned} Z_1 &= (1/R_1 + 1/R_A + 1/R_B)^{-1} \\ Z_2 &= (1/R_C + 1/R_2 + 1/Z)^{-1} \end{aligned} \tag{5.155}$$

In most cases of practical interest, Z_1 and R_1 are nearly equal. Thus, G becomes

$$G \doteq -\frac{Z_2 h_{fe}}{R_1 + h_{ie}} \tag{5.156}$$

Let us now assume the resonator Z in Figure 5.25 is of the energy trapping type. If the electrodes are small enough to eliminate all the spurious members of the anharmonic series, Z_2 may be characterized in terms of the equivalent circuit parameters appearing in Figure 5.22 as

$$Z_2 = \frac{R_p A(\omega)}{A(\omega) + j\omega R_p(C_S + C_{11})(1 - \omega^2 L_{11} C_s)} \tag{5.157}$$

where

$$R_p = (1/R_C + 1/R_2)^{-1} \tag{5.158}$$

$$A(\omega) = 1 - \omega^2 L_{11} C_{11} \tag{5.159}$$

$$C_s = (1/C_S + 1/C_{11})^{-1} \tag{5.160}$$

Note that the only frequency dependence of G in Equation 5.156 is that of the Z_2 factor which is expressed by Equations 5.157 to 5.160. For large R_p, consequently, G is a maximum near the resonator's antiresonance

$$\omega_a = \frac{1}{\sqrt{L_{11} C_s}} \tag{5.161}$$

The resistance R_p may be regarded as the source impedance through which the resonator is driven. Consequently, Equation 5.161 illustrates the concept discussed in Section 5.4.1: The frequency at which system response is maximum will approach the resonator antiresonance as driving source impedance becomes large.

Equation 5.161 determines the frequency of maximum G for the circuit in Figure 5.25. It is also useful to find the $3dB$ bandwidth of the G maximum. Let us express as $\Delta\omega$ the separation between the resonator resonance and antiresonance:

$$\Delta\omega = \omega_a - \omega_r = \frac{1}{\sqrt{L_{11} C_s}} - \frac{1}{\sqrt{L_{11} C_{11}}} \tag{5.162}$$

Furthermore, let the $3dB$ bandwidth be written as $b = 2\delta\omega$. Then, if $\delta\omega$ is significantly smaller than $\Delta\omega$, the $3dB$ points $\omega = \omega_a \pm \delta\omega$ are those frequencies at which

$$A(\omega_a) = |\omega_a R_p(C_S + C_{11})(1 - \omega^2/\omega_a^2)| \tag{5.163}$$

If we substitute $\omega = \omega_a + \delta\omega$ in Equation 5.163, we can find

$$A(\omega_a) = \frac{\omega_a R_p(C_S + C_{11})}{Q} \tag{5.164}$$

where Q is the quality factor of the gain maximum at ω_a,

$$Q = \frac{\omega_a}{2\delta\omega} = \frac{\omega_a}{b} \tag{5.165}$$

On the other hand, $A(\omega_a)$ may also be evaluated from Equations 5.159 and 5.162 as

$$A(\omega_a) = \frac{2\Delta\omega}{\omega_a} \tag{5.166}$$

Consequently, the $3dB$ bandwidth becomes

$$b = \frac{2\Delta\omega}{R_p(C_S + C_{11})\omega_a} \doteq \frac{2\Delta\omega}{R_p C_S \omega_a} \tag{5.167}$$

Let us now consider the problem of designing a 50 Ω output impedance PZT-5A filter of center frequency $f_a = 10.7$ Mc and of 4% bandwidth. Let us do this by utilizing a thickness-dilational trapped-energy resonator with circular electrodes, as described in Section 5.3.

This filter, first of all, requires $\delta\omega = 0.02\ \omega_a$. We need now to check that the requirement is met of $\delta\omega$ being significantly smaller than $\Delta\omega$. From Figure 5.23, we see that C_{11N} is about 0.6 over most values of electrode diameter to plate thickness ratio. In view of Equation 5.150, this means,

$$\frac{C_S}{C_{11}} = \left(\frac{k_t^2}{1 - k_t^2} C_{11N}\right)^{-1} = 8.508 \tag{5.168}$$

Here the k_t value appropriate to PZT-5A has been substituted (see Table 5.1 and Equation 5.104). Equations 5.160, 5.162, and 5.168 may then be combined to give $\Delta\omega = 0.0541\ \omega_a$. Consequently, $\delta\omega$ is indeed significantly smaller than $\Delta\omega$.

Let us next estimate the half-thickness h of the required ceramic plate. We want the system resonance or G maximum to be at 10.7 Mc. Thus the resonator antiresonance $\omega_a = 1.0571\ \omega_r$ should occur at this same point. On the other hand, the resonator resonance ω_r will be very slightly above the electroded region cutoff frequency, $\Omega_{co(2)} = 2.2658$. Equation

5.87 then yields the desired plate half-thickness as

$$h = \frac{\Omega_{co(2)}\pi}{2\omega_r}\sqrt{\frac{c_{55}^E}{\rho}}$$

$$= \frac{\Omega_{co(2)}\pi}{2\omega_a}\cdot 1.0571\sqrt{\frac{c_{55}^E}{\rho}} \qquad (5.169)$$

When $2\pi \times 10.7 \times 10^6$ is substituted for ω_a, and when ρ and c_{55}^E are substituted from Table 5.1, h comes out to be 0.0985 mm.

The required electrode radius a is now determined from Equation 5.167:

$$C_S = \frac{\pi a^2 \epsilon_{33}^S}{2h} = \frac{2\Delta\omega}{bR_p\omega_a} \qquad (5.170)$$

We solve this equation for a. Then we substitute ϵ_{33}^S from Table 5.1 and let $h = 0.985 \times 10^{-4}$, $\Delta\omega = 0.054\,\omega_a$, $b = 0.04\,\omega_a$, where $\omega_a = 2\pi \times 10.7 \times 10^6$. Also R_p is 50 Ω, as the circuit is specified to have a 50 Ω output impedance. These values lead to the result $a =$ 2.604 mm.

The a and h dimensions just evaluated yield the $a/h = B$ ratio of 26.5. Reference to Figure 5.23 then indicates that this ratio is too large to give a single clean response. Rather, there will be three closely spaced responses characterized by $C_{1nN} = 0.599$, 0.091, and 0.011. The associated normalized resonant frequencies are

$$(\Omega_{1n} - \Omega_{co(2)})/(\Omega'_{co(2)}[c'_{55}/c_{55}^E]^{1/2} - \Omega_{co(2)}) = 0.092, 0.444, \text{ and } 0.921$$

respectively.

If we compute G as a function of ω from Equations 5.155 and 5.156 with the effects of C_{12N} and C_{13N} neglected in Z,

$$Z = \left(j\omega C_S + \frac{j\omega C_{11}}{1 - \omega^2/\omega_{11}^2}\right)^{-1} \qquad (5.171)$$

we get the normalized response curve shown in Figure 5.26 by the dashed line. Note that G is indeed maximum around 10.7 Mc, and that b is 0.048 ω_a. The slight bandwidth broadening over the desired 0.040 ω_a occurs because an elastic loss tangent of 1/150 and a dielectric loss tangent of 1/100 are assumed in Figure 5.26. The theoretical bandwidth computed in Equation 5.167, on the other hand, assumes no material losses.

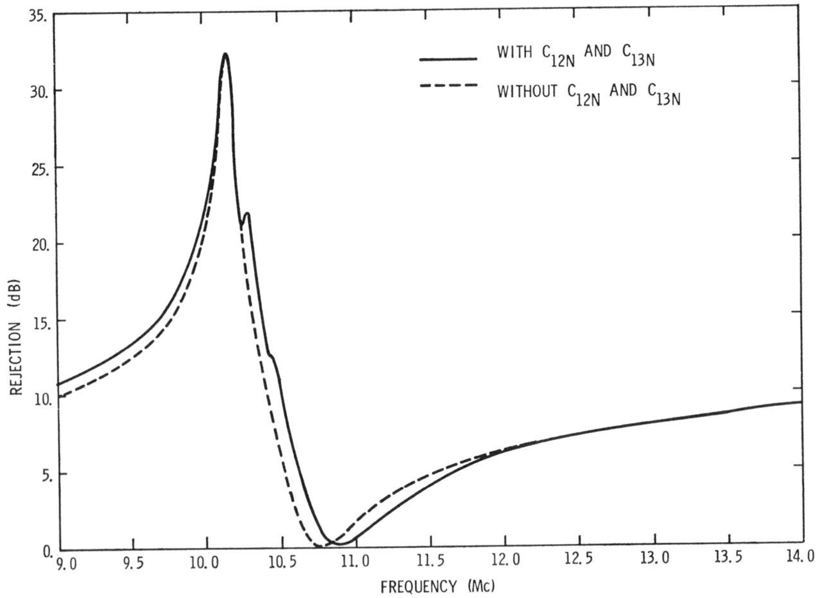

Figure 5.26. Normalized response curves for the filter circuit shown in Figure 5.25. These curves are based on a PZT-5A trapped energy resonator with circular electrodes of 2.604 mm radius on a plate 0.197 mm thick. One curve includes only the effect of C_{11N}, while the other curve includes the effects of C_{12N} and C_{13N} as well. Note that the inclusion of C_{12N} and C_{13N} gives rise to two small spurious responses on the *low* frequency side of the gain maximum.

If G is determined as a function of ω with the effects of C_{12N} and C_{13N} included,

$$Z = \left(j\omega C_S + \sum_{n=1}^{3} \frac{j\omega C_{1n}}{1 - \omega^2/\omega_{1n}^2}\right)^{-1} \tag{5.172}$$

we get the normalized response curve shown by the solid line in Figure 5.26. It may be seen on this curve that the third resonance greatly dominates the first two. The first two resonances appear as weak perturbations on the *lower* side of the G maximum. For that reason, the casual observer may not realize they are spurious members of the anharmonic series.

References

1. D. R. Curran and W. J. Gerber, "Low-Frequency Ceramic Band-Pass Filters" *Proc. 1963 Electronics Components Conference*, Washington, D.C., May 7–9, 1963.
2. D. J. Koneval, W. J. Gerber, and D. R. Curran, "Improved VHF Filter Crystals Using Insulating Film Techniques," *Proc. 20th Annual Symposium on Frequency Control*, pp. 103–130 (April 1966).
3. E. G. Spencer, P. V. Lenzo, and A. A. Ballman, "Dielectric Materials for Electro-optic, Elastooptic, and Ultrasonic Device Applications," *Proc. IEEE*, **55**, pp. 2074–2108 (December 1967).
4. C. E. Land, G. W. Smith, and I. D. McKinney, "Polycrystalline Ferroelectric Multiremanence Memory Elements," *1964 IEEE International Convention Rec.*, **12,** pt. 9, pp. 149–160.
5. D. G. Schueler, "Ferroelectric Ceramic Logic and NDRO Memory Devices," *1966 WESCON Convention Rec.*, **10,** pt. 3, paper 3/4.
6. C. E. Land, "Small-Signal Applications of Monolithic Multiport Piezoelectric Devices," *1966 WESCON Convention Rec.*, **10,** pt. 3, paper 3/5.
7. A. Benjaminson, "The Linear Quartz Thermometer—a New Tool for Measuring Absolute and Difference Temperatures," *Hewlett-Packard J.*, **16,** No. 7, pp. 1–7 (March 1965).
8. K. Blötekjaer and C. F. Quate, "The Coupled Modes of Acoustic Waves and Drifting Carriers in Piezoelectric Crystals," *Proc. IEEE*, **52,** pp. 360–377 (April 1964).
9. D. L. White, "Amplification of Ultrasonic Waves in Piezoelectric Semiconductors," *J. Appl. Phys.*, **33,** pp. 2547–2554 (August 1962).
10. W. E. Newell, "Tuned Integrated Circuits—a State-of-the-Art Survey," *Proc. IEEE*, **52,** pp. 1603–1608 (December 1964).
11. "Standards on Piezoelectric Crystals, 1949," *Proc. IRE*, **37,** pp. 1378–1395 (December 1949).
12. D. A. Berlincourt, D. R. Curran, and H. Jaffe, "Piezoelectric and Piezomagnetic Materials and Their Function in Transducers," in *Physical Acoustics*, W. P. Mason, Ed., **1,** pt. A. Academic Press, Inc., New York, 1964, pp. 182–189.

13. R. N. Thurston, "Wave Propagation in Fluids and Normal Solids," in *Physical Acoustics*, W. P. Mason, Ed., **1**, pt. A. Academic Press, Inc., New York, 1964, pp. 39–41.
14. Reference 12, p. 171.
15. K. S. Van Dyke and E. S. Creeger, "Piezoelectric Crystal Studies and Measurements," *First Quart. Prog. Rept. of Wesleyan University* (February 1 to April 30, 1953) U.S. Army Signal Corps Contract No. DA-36-039 SC-42587, June 1953.
16. R. Bechmann, "Elastic and Piezoelectric Constants of Alpha-Quartz," *Phys. Rev.*, **110**, pp. 1060–1061 (June 1958).
17. Reference 12, pp. 202–204.
18. A. W. Warner, M. Onoe, and G. A. Coquin, "Determination of Elastic and Piezoelectric Constants for Crystals in Class (3m)," *J. Acoust. Soc. Am.*, **42**, pp. 1223–1231 (December 1967).
19. M. Onoe, A. W. Warner, and A. A. Ballman, "Elastic and Piezoelectric Characteristics of Bismuth Germanium Oxide $Bi_{12}GeO_{20}$," *IEEE Trans. Sonics and Ultrasonics*, **SU-14**, pp. 165–167 (October 1967).
20. A. A. Ballman, "The Growth and Properties of Piezoelectric Bismuth Germanium Oxide $Bi_{12}GeO_{20}$," *J. Crystal Growth*, **1**, pp. 37–40 (January 1967).
21. G. E. Martin, "Vibrations of Longitudinally Polarized Ferroelectric Cylindrical Tubes," *J. Acoust. Soc. Am.*, **35**, pp. 510–520 (April 1963).
22. G. E. Martin, "Dielectric, Piezoelectric, and Elastic Losses in Longitudinally Polarized Segmented Ceramic Tubes," *U.S. Navy J. Underwater Acoustics*, **15**, pp. 329–332 (April 1965).
23. C. E. Land, G. W. Smith, and C. R. Westgate, "The Dependence of the Small-Signal Parameters of Ferroelectric Ceramic Resonators upon State of Polarization," *IEEE Trans. Sonics and Ultrasonics*, **SU-11**, pp. 8–19 (June 1964).
24. P. M. Morse and H. Feshbach, *Methods of Theoretical Physics*, pt. I. McGraw-Hill Book Company, Inc., New York, 1953, pp. 114–115.
25. F. B. Hildebrand, *Methods of Applied Mathematics*. Prentice-Hall Publishing Company, Inc., Englewood Cliffs, N.J., 1952, p. 46.
26. Ibid., pp. 49–53.
27. R. Holland, *Application of Green's Functions and Eigenmodes in the Design of Piezoelectric Ceramic Devices*, Ph.D. Thesis, Electrical Engineering, Massachusetts Institute of Technology, January 1966, pp. 257–266, 286–304.
28. R. Bechmann, "Some Applications of the Linear Piezoelectric Equations of State," *IRE Trans. Ultrasonics Eng.*, **UE-3**, pp. 43–62 (May 1955).
29. Reference 12, pp. 190–193.
30. H. G. Baerwald, "Eigen Coupling Factors and Principal Components, the Thermodynamic Invariants of Piezoelectricity," *1960 IRE International Convention Rec.*, **8**, pt. 6, pp. 205–211.
31. H. G. Baerwald, "The Invariant Normal Representation of Twin-Property Thermodynamic Potentials with Application to Piezoelectricity," Sandia Corporation, Albuquerque, New Mexico, Research Report, SC-RR-68-213, April 1968.
32. Reference 25, pp. 168–176.
33. Ibid., p. 23.
34. W. P. Mason, "Use of Piezoelectric Crystals and Mechanical Resonators in Filters and Oscillators," in *Physical Acoustics*, W. P. Mason, Ed., **1**, pt. A. Academic Press, Inc., New York, 1964, pp. 336–355.
35. J. A. Lewis, "The Effect of Driving Electrode Shape on the Electrical Properties of Piezoelectric Crystals," *The Bell System Tech. J.*, **40**, pp. 1259–1280 (September 1961).
36. R. C. Buck, *Advanced Calculus*. McGraw-Hill Book Company, Inc., New York, 1956, p. 74 (problem 14).

37. P. R. Halmos, *Finite-Dimensional Vector Spaces.* D. Van Nostrand Company, Inc., Princeton, N.J., 1958, pp. 98–101.
38. Reference 13, pp. 56–58.
39. J. D. Jackson, *Classical Electrodynamics.* John Wiley & Sons, Inc., New York, 1962, p. 256.
40. W. P. Mason, "Electrostrictive Effect in Barium Titanate Ceramics," *Phys. Rev.*, **74,** pp. 1134–1147 (November 1948).
41. Reference 12, pp. 219–233.
42. R. Holland, "The Equivalent Circuit of an N-Electrode Piezoelectric Bar," *Proc. IEEE,* **54,** pp. 968–975 (July 1966).
43. R. Holland, "The Equivalent Circuit of a Symmetric N-Electrode Piezoelectric Disk," *IEEE Trans. Sonics and Ultrasonics,* **SU-14,** pp. 21–33 (January 1967).
44. B. van der Veen, "The Equivalent Network of a Piezoelectric Crystal with Divided Electrodes," *Philips Res. Repts.*, **11,** pp. 66–79 (February 1956).
45. E. C. Munk, "The Equivalent Electrical Circuit for Radial Modes of a Piezoelectric Ceramic Disc with Concentric Electrodes," *Philips Res. Repts.*, **20,** pp. 170–189 (April 1965).
46. Reference 24, pp. 791–895.
47. Reference 25, pp. 388–406.
48. E. H. Jacobsen, "Sources of Sound in Piezoelectric Crystals," *J. Acoust. Soc. Am.*, **32,** pp. 949–953 (August 1960).
49. D. Berlincourt, "Variation of Electroelastic Constants of Polycrystalline Lead Titanate Zirconate with Thoroughness of Poling," *J. Acoust. Soc. Am.*, **36,** pp. 515–520 (March 1964).
50. Reference 24, pp. 804–810.
51. C. E. Land, "Transistor Oscillators Employing Piezoelectric Ceramic Feedback Networks," *1965 IEEE International Convention Rec.*, **13,** pt. 7, pp. 51-68.
52. Reference 27, pp. 75–77.
53. Reference 24, pp. 719-757.
54. Reference 25, pp. 95–100, pp. 144–147.
55. A. E. H. Love, *A Treatise on the Mathematical Theory of Elasticity.* Dover Publications, New York, 1944, pp. 180–181.
56. M. Onoe and T. Kurachi, "Non-Axisymmetric Vibrations of a Circular Piezoelectric Ceramic Disk," *Electronics and Communications (Japan),* **49,** No. 1, pp. 67–75 (January 1966).
57. R. Holland, "Numerical Studies of Elastic-Disk Contour Modes Lacking Axial Symmetry," *J. Acoust. Soc. Am.*, **40,** pp. 1051–1057 (November 1966).
58. M. Onoe, "Contour Vibrations of Isotropic Circular Plates," *J. Acoust. Soc. Am.*, **28,** pp. 1158–1162 (November 1956).
59. Reference 55, pp. 497–498.
60. Reference 27, pp. 124–126.
61. R. Holland, "Analysis of Multiterminal Piezoelectric Plates," *J. Acoust.* Soc. *Am.*, **41,** pp. 940–952 (April 1967).
62. W. E. Newell, "Ultrasonics in Integrated Electronics," *Proc. IEEE,* **53,** pp. 1305–1309 (October 1965).
63. W. E. Newell, "Face-Mounted Piezoelectric Resonators," *Proc. IEEE,* **53,** pp. 575–581 (June 1965).
64. G. E. Martin, "On the Theory of Segmented, Electromechanical Systems," *J. Acoust. Soc. Am.*, **36,** pp. 1366–1370 (July 1964).
65. G. E. Martin, "Vibrations of Coaxially Segmented, Longitudinally Polarized Ferroelectric Tubes," *J. Acoust. Soc Am.*, **36,** pp. 1496–1506 (August 1964).
66. D. K. Winslow and H. J. Shaw, "Multiple Film Microwave Acoustic Transducers," *1966 IEEE International Convention Rec.*, **14,** pt. 5, pp. 26–31.

67. E. K. Sittig, "Transmission Parameters of Thickness-Driven Piezoelectric Transducers Arranged in Multilayer Configurations," *IEEE Trans. Sonics and Ultrasonics*, **SU-14,** pp. 167–174 (October 1967).
68. R. Holland, "The Linear Theory of Multielectrode Piezoelectric Plates," *1966 WESCON Convention Rec.*, **10,** pt. 3, paper 3/3.
69. E. P. EerNisse, "Resonances of One-Dimensional Composite Piezoelectric and Elastic Structures," *IEEE Trans. Sonics and Ultrasonics*, **SU-14,** pp. 59–67. (April 1967).
70. Lord Rayleigh, *The Theory of Sound*, **1**. Dover Publications, New York, 1945, pp. 103–112.
71. Reference 55, pp. 166–169.
72. H. Ekstein, "Free Vibrations of Anisotropic Bodies," *Phys. Rev.*, **66,** pp. 108–118 (September 1944).
73. Reference 25, pp. 191–193, pp. 224–225 (problem 81).
74. S. H. Gould, *Variational Methods for Eigenvalue Problems*. University of Toronto Press, Toronto, and Oxford University Press, London, 1966, pp. 91–92.
75. Reference 24, pp. 44–50.
76. Ibid., p. 116.
77. W. G. Bickley and R. E. Gibson, *Via Vector to Tensor*. John Wiley & Sons, Inc., New York, 1962, p. 70.
78. E. P. EerNisse, "Computer Program for Calculating the Resonant Properties of Piezoelectric Ceramic Discs with Comparable Diameter and Thickness Dimensions," Sandia Corporation, Albuquerque, New Mexico, Research Report SC-RR-66-550, October 1966.
79. E. A. G. Shaw, "On the Resonant Vibrations of Thick Barium Titanate Discs," *J. Acoust. Soc. Am.*, **28,** pp. 38–50 (January 1956).
80. D. C. Gazis and R. D. Mindlin, "Extensional Vibrations and Waves in a Circular Disc and a Semi-Infinite Plate," *J. Appl. Mech., Trans. ASME*, **27,** pp. 541–547 (September 1960).
81. R. Holland, "Contour Extensional Resonant Properties of Rectangular Piezoelectric Plates," *IEEE Trans. Sonics and Ultrasonics*, **SU-15,** pp. 97–105 (April 1968).
82. R. Holland and A. L. Roark, "Computer Program for Analyzing Free, Rectangular Piezoelectric Plates," Sandia Corporation, Albuquerque, New Mexico, Research Report SC-RR-67-524, August 1967.
83. P. Lloyd and M. Redwood, "Finite-Difference Method for the Investigation of the Vibrations of Solids and the Evaluation of the Equivalent-Circuit Characteristics of Piezoelectric Resonators, Parts I and II," *J. Acoust. Soc. Am.*, **39** pp. 346–361 (February 1966).
84. H. Mähly and A. Trösch, "Shear Modes of Square Plates," *Helv. Phys. Acta*, **20,** pp. 253–255 (August 1947).
85. M. G. Lamé, *Leçons sur la Théorie Mathématique de l'Élasitcité des Corps Solides*. Bachelier, Paris, 1852, pp. 165–172.
86. V. Petrzilka, "Longitudinal Vibrations of Rectangular Quartz Plates," *Z. Physik*, **47,** pp. 436–454 (October 1935).
87. L. Collatz, "Approximate Computation of Eigenvalues," *Z. Angew. Math. Mech.*, **19,** pp. 224–249 (August 1939).
88. Reference 74, pp. 31–40.
89. H. Mähly, "Eigenmodes of Thin Square Crystal Plates," *Helv. Phys. Acta*, **18,** pp. 248–251 (July 1945).
90. H. G. Baerwald and C. Libove, "Breathing Vibrations of Planarly Isotropic Square Plates," Clevite Research Center, Cleveland, Ohio, Contract Nonr-1055(00), Tech. Rept. #8, December 1955. Many of the results of this report, which is not generally available, are reproduced in Reference 91.

91. "IRE Standards on Piezoelectric Crystals: Determination of the Elastic, Piezoelectric, and Dielectric Constants—the Electromechanical Coupling Factor, 1958," *Proc. IRE*, **46**, pp. 764–778 (April 1958).
92. M. Onoe, "Contour Vibrations of Thin Rectangular Plates," *J. Acoust. Soc. Am.*, **30**, pp. 1159–1162 (December 1958).
93. M. A. Medick and Y. H. Pao, "Extensional Vibrations of Thin Rectangular Plates," *J. Acoust. Soc. Am.*, **37**, pp. 59–65 (January 1965).
94. M. Onoe and Y. H. Pao, "Edge Mode of Thin Rectangular Plate of Barium Titanate," *J. Acoust. Soc. Am.*, **33**, p. 1628 (November 1961).
95. E. P. EerNisse, "Coupled-Mode Approach to Elastic-Vibration Analysis," *J. Acoust. Soc. Am.*, **40**, pp. 1045–1050 (November 1966).
96. R. Holland and E. P. EerNisse, "Variational Evaluation of Admittances of Multielectroded Three-Dimensional Piezoelectric Structures," *IEEE Trans. Sonics and Ultrasonics*, **SU-15**, pp. 119–132 (April 1968).
97. E. P. EerNisse, "Variational Method for Electroelastic Vibration Analysis," *IEEE Trans. Sonics and Ultrasonics*, **SU-14**, pp. 153–160 (October 1967).
98. H. F. Tiersten, "Natural Boundary and Initial Conditions from a Modification of Hamilton's Principle," *J. Math. Phys.*, **9**, pp. 1445–1450 (September 1968).
99. P. Lloyd, "Equations Governing the Electrical Behavior of an Arbitrary Piezoelectric Resonator Having N Electrodes," *Bell System Tech. J.*, **46**, pp. 1881–1900 (October 1967).
100. H. Ekstein and T. Schiffman, "Free Vibrations of Isotropic Cubes and Nearly Cubic Parallelepipeds," *J. Appl. Phys.*, **27**, pp. 405–412 (April 1956).
101. H. Ekstein and T. Schiffman, "Elastic Vibrations of Almost Cubic Parellelepipeds," *J. Appl. Phys.*, **22**, p. 1215 (September 1951).
102. R. Holland, "Computer Program for Analyzing Free Rectangular Piezoelectric Parallelepipeds," Sandia Corporation, Albuquerque, New Mexico, Research Report, to be published.
103. V. Heine, *Group Theory in Quantum Mechanics*. Pergamon Press, New York, 1960, pp. 222–224.
104. H. Eyring, J. Walter, and G. E. Kimball, *Quantum Chemistry*. John Wiley & Sons, Inc., New York, 1944, pp. 172–189, pp. 376–388.
105. Reference 103, pp. 448–454.
106. F. A. Cotton, *Chemical Applications of Group Theory*, Appendix II. Interscience Publishers, Inc., New York, 1963, p. 9.
107. Reference 12, p. 229.
108. R. Holland, "Resonant Properties of Piezoelectric Ceramic Rectangular Parallelepipeds," *J. Acoust. Soc. Am.*, **43**, pp. 988–997 (May 1968).
109. W. S. Mortley, "F.M.Q.," *Wireless World*, **57**, pp. 399–403 (October 1951).
110. W. S. Mortley, "Frequency-Modulated Quartz Oscillators for Broadcasting Equipment," *Proc. IEE* (*London*), **104**, part B, paper 2302R, pp. 239–253 (May 1957).
111. W. Shockley, D. R. Curran, and D. J. Koneval, "Energy Trapping and Related Studies of Multiple Electrode Filter Crystals," *Proc. 17th Annual Symposium on Frequency Control*, pp. 88–126 (April 1963).
112. W. H. Horton and R. C. Smythe, "The Work of Mortley and the Energy-Trapping Theory for Thickness-Shear Piezoelectric Vibrators," *Proc. IEEE*, **55**, p. 222 (February 1967).
113. R. Bechmann, U.S. Patent No. 2,249,933, 1941.
114. R. Bechmann, "Quartz AT-Type Filter Crystals for the Frequency Range 0.7 to 60 Mc," *Proc. IEEE*, **49**, pp. 523–524 (February 1961).
115. E. A. Gerber, "Comments on Unwanted Responses in VHF Crystals," *Proc. 20th Annual Symposium on Frequency Control*, pp. 161–166 (April 1966).

116. H. Mailer and D. R. Beuerle, "Incorporation of Multi-Resonator Crystals into Filters for Quantity Production," *Proc. 20th Annual Symposium on Frequency Control*, pp. 309–342 (April 1966).
117. M. Onoe and H. Jumonji, "Analysis of Piezoelectric Resonators Vibrating in Trapped-Energy Modes," *Electronics and Communications* (*Japan*), **48,** No. 9, pp. 84–93 (September 1965).
118. R. A. Sykes, W. L. Smith, and W. J. Spencer, "Monolithic Crystal Filters," *1967 IEEE International Convention Rec.*, **15,** pt. 11, pp. 78–93.
119. R. A. Sykes and W. D. Beaver, "High Frequency Monolithic Crystal Filters with Possible Applications to Single Frequency and Single Side Band Use," *Proc. 20th Annual Symposium on Frequency Control*, pp. 288–308 (April 1966).
120. M. Onoe, H. Jumonji, and N. Kobori, "High Frequency Crystal Filters Employing Multiple Mode Resonators Vibrating in Trapped Energy Modes," *Proc. 20th Annual Symposium on Frequency Control*, pp. 266–287 (April 1966).
121. A. Ballato, T. Lukaszek, H. Wasshausen, and E. Chabak, "Design and Fabrication of Modern Filter Crystals," *Proc. 20th Annual Symposium on Frequency Control*, pp. 131–160 (April 1966).
122. D. R. Curran, D. J. Koneval, R. B. McEntee, and K. A. Pim, "Research into New Approaches for VHF Filter Crystals," Sixth Summary Report (August 1, 1964 to January 15, 1965) of Clevite Corp., Electronic Research Div., Cleveland, Ohio, U.S. Army Electronics Laboratory Contract No. DA 36-039 AMC-02245(E), April 1965.
123. T. Kaname, T. Nagata, and T. Nakajima, "Piezoelectric Ceramic Trapped Energy Resonators," *Proceedings of the Barium Titanate Study Meeting of Japan*, pp. 31–33 (May 1966). English Translation: SC-T-68-1544, Sandia Corporation, Albuquerque, New Mexico.
124. M. de Jong, Private communication, N. V. Philips, Eindhoven, Netherlands.
125. R. D. Mindlin and P. C. Y. Lee "Thickness-Shear and Flexural Vibrations of Partially Plated, Crystal Plates," *International Journal of Solids and Structures*, **2,** pp. 125–139 (January 1966).
126. R. D. Mindlin and M. A. Medick "Extensional Vibrations of Elastic Plates," *J. App. Mech., Trans. ASME*, **26,** pp. 561–569 (December 1959).
127. H. F. Tiersten and R. D. Mindlin, "Forced Vibrations of Piezoelectric Crystal Plates," *Quart. Appl. Math.*, **20,** pp. 107–119 (July 1962).
128. H. F. Tiersten, "Wave Propagation in an Infinite Piezoelectric Plate," *J. Acoust. Soc. Am.*, **35,** pp. 234–239 (February 1963).
129. R. D. Mindlin, "An Introduction to the Mathematical Theory of Vibrations of Elastic Plates," U.S. Army Signal Corps Contract No. DA-36-039 SC-56772, Fort Monmouth, N.J., 1955, p. 2–43.
130. R. Holland and E. P. EerNisse, "Accurate Measurement of Coefficients in a Ferroelectric Ceramic," *IEEE Trans. Sonics and Ultrasonics*, **SU-19,** October 1969.
131. "IRE Standards on Piezoelectric Crystals: Measurements of Piezoelectric Ceramics, 1961," *Proc. IRE*, **49,** pp. 1161–1169 (July 1961).
132. J. A. Straton, *Electromagnetic Theory*. McGraw-Hill Book Company, Inc., New York, 1941, pp. 355–391.
133. F. Sauerland and W. Blum, "Ceramic IF Filters for Consumer Products," *IEEE Spectrum*, **5,** No. 11, pp. 112–126 (November 1968).

Index